977.43
6/26/03

Location Management and Routing in Mobile Wireless Networks

For a complete listing of the *Artech House Mobile Communications Library*, turn to the back of this book.

Location Management and Routing in Mobile Wireless Networks

Amitava Mukherjee
Somprakash Bandyopadhyay
Debashis Saha

Artech House
Boston • London
www.artechhouse.com

Library of Congress Cataloging-in-Publication Data
Mukherjee, Amitava, 1959–
    Location management and routing in mobile wireless networks / Amitava Mukherjee,
Somprakash Bandyopadhyay, Debashis Saha.
    p. cm. — (Artech House mobile communications series)
    Includes bibliographical references and index.
    ISBN 1-58053-355-8 (alk. paper)
    1. Wireless communication systems—Location.   2. Wireless communication systems—
Management.   3. Routers (Computer networks) I. Bandyopadhyay, Somprakash, 1957–
II. Saha, Debashis, 1965–   III. Title.   IV. Series.
TK5103.2.M85 2003
621.382—dc21
                                                  2003041889

British Library Cataloguing in Publication Data
Mukherjee, Amitava, 1959–
Location management and routing in mobile wireless networks. — (Artech House mobile com-
munications series)
    1. Mobile communication systems   I. Title   II. Bandyopadhyay, Somprakash, 1957–
III. Saha, Debashis, 1965–
    621 . 3'8456

    ISBN 1-58053-355-8

Cover design by Yekaterina Ratner

© 2003 ARTECH HOUSE, INC.
685 Canton Street
Norwood, MA 02062

International Standard Book Number: 1-58053-355-8
Library of Congress Catalog Card Number: 2003041889

10 9 8 7 6 5 4 3 2 1

Contents

Preface

This book aims at presenting, in a canonical form, the work done by us in the field of routing in mobile wireless networks. Most of the material contained herein has previously been presented at international conferences or has been accepted for publication in journals.

Mobile wireless networks can be broadly classified into two distinct categories: infrastructured (cellular) and infrastructureless (ad hoc). While cellular networks usually involve a single-hop wireless link to reach a mobile terminal, ad hoc networks normally require a multihop wireless path from a source to a destination. The growth of mobility aspects in cellular networks is occurring at three different levels. First, growth occurs at the spatial level (i.e., users desire to roam with a mobile terminal). Second, growth occurs from the penetration rate of mobile radio access lines. And third, the traffic generated by each wireless user is constantly growing. On one hand, tetherless (e.g., cellular) subscribers use their mobile terminals; on the other hand, the arrival of more capacity-greedy services (e.g., Internet accesses, multimedia services). From all of these considerations, the generalized mobility features will have serious impacts on the wireless telecommunications networks. Mobility can be categorized into two areas: radio mobility, which mainly consists of the handover process and network mobility, which mainly consists of location management (location updating and paging). In this book, we shall concentrate on the network mobility only.

This book will act as a general introduction to location management, and routing in both single-hop and multihop mobile wireless networks, so that readers can gain familiarity with location management and routing issues in this field. In particular, it will provide the details of location management and paging in wireless cellular networks, and routing in mobile ad hoc networks.

In about 200 pages, it will cover the past, present, and future works on location management and routing protocols in all types of mobile wireless networks. In cellular networks, the emphasis will be on mobility issues, location management, paging, and radio resources. In mobile ad hoc networks, the focus will be on different types of routing protocols and medium access control techniques. It will discuss numerous potential applications, review relevant concepts, and examine the various approaches that enable readers to understand the issues and future research problems in this field too. In a word, it will cover everything you can think in the realm of location management and routing issues in mobile wireless networks.

Barring Chapter 1, which is a general introduction to the subject, the book is divided into three parts, namely Part I: Cellular Networks, Part II: Ad Hoc Wireless Networks, and Part III: Future Issues. In Part I, there are three chapters. Chapter 2 concentrates on two important mobility issues, namely mobility models (fluid-flow model, random walk model, gravity model), and mobility traces (metropolitan mobility, national mobility, international mobility). Chapter 3 concerns radio resource management, including radio propagation, and channel assignment. Chapter 4 describes an important issue called location management. It covers issues such as paging (blanket paging, and intelligent paging), location update (static location update, dynamic location update) and location area planning (manual registration, automatic location management using location area, memory-based location management methods, non-memory-based location methods, location management in CDPD, GPRS, WCDMA, and IMT-2000).

Part II focuses on ad hoc wireless networks and again comprises three chapters. Chapter 5 is an overview of the characteristics of ad hoc networks including three fundamental design choices, namely flat versus hierarchical architecture, proactive versus reactive routing, and medium access protocols. Chapter 6 describes medium access control techniques in detail, covering basic media access protocol for wireless LANs (IEEE 802.11), Floor Acquisition Multiple Access, Dual Busy Tone Multiple Access, Power Controlled Multiple Access Protocols, MAC with Adaptive Antenna, Directional MAC Protocols, and Adaptive MAC Protocol for WACNet. Chapter 7 discusses both unicast and multicast routing protocols in ad hoc wireless networks. Unicast routing techniques include proactive routing protocols, such as DSDV, WRP, CBR, CGSR, OLSR, FSR, and agent-based protocols for topology discovery and routing, and reactive routing protocols, such as DSR, AODV, TORA, ABR, SSA, stability-based routing, LAR, and query localization techniques for on-demand routing. It also includes power-aware routing, multipath routing, and QoS Management.

Part III explores future issues such as routing in next-generation wireless networks, location management in all-IP IMT-2000 networks, routing in ad hoc sensor networks, and routing in pervasive networks.

This book is a uniquely comprehensive study of the major location management and routing technologies and systems that will assist in forming the future mobile wireless networks. We have written the book for those professionals and students who want such a comprehensive view. It may be used as a text or reference book in graduate courses in mobile wireless networks.

Acknowledgments

We want to express our sincere gratitude towards Artech House Books for giving us the opportunity to write on this topic. Many thanks also to our colleagues at the department, past and present, for many rewarding discussions and for contributing to the stimulating and pleasant atmosphere.

Several other people helped us during the course of writing this book. We would like to specially thank our colleagues at Indian Institute of Management (IIM), Calcutta and *i*-SDC SBU, IBM Global Services, Calcutta. Special thanks go to Amitabh Ray, Agnimitra Biswas, Surojit Mookherjee and Reena J. Sarkar of IBM Global Services, Calcutta and Jaydeep Mukherjee of Cogentech Management Consultants (P) Ltd., Calcutta.

The main bulk of the work was carried out by our doctoral students, namely Partha Sarathi Bhattacharjee of Bharat Sanchar Nigam Ltd., Calcutta and Krishna Paul of Indian Institute of Technology, Bombay. We express our gratitude to them. Many thanks to Sauti Sen for designing the cover layout.

It is with pleasure that we also acknowledge and thank the editorial staff of Artech House Books. Tiina Ruonamaa, assistant editor, and Dr. Julie Lancashire, senior commissioning editor, Artech House Books, have helped with logistics and with their enthusiasm in giving the prompt reminders before the promised deadlines. Finally, our thanks go to the production department of Artech House Books for managing with a very tight schedule.

Last but not least, we want to thank our families for their support and encouragement throughout this time.

Part I
Cellular Networks

1

Introduction

Wireless communication has recently captured the attention and the imagination of users from all walks of life. The major goal of wireless communication is now to allow a user to have access to the capabilities of global networks at any time without regard to location or mobility. Since their emergence in the 1970s [1], the mobile wireless networks have become increasingly popular in the networking industry. This has been particularly true within the past decade, which has seen wireless networks being adapted to enable mobility. Since the inception of cellular telephones in the early 1980s [2], they have evolved from a costly service with limited availability toward an affordable and more versatile alternative to wired telephony. In the future, it appears that, not only will cellular installations continue to proliferate, but wireless access to fixed telephones will become much more common.

Trends in wireless communication are proceeding with a strong tendency toward increasing need for mobility in the access links within the network. Examples are (1) residential line access with the proliferation of cordless phones and their penetration rate having passed that of fixed phones in several countries including the United States and Japan; (2) business lines with wireless private branch exchange (WPBX) access for voice services, and wireless LANs (WLANs) for computer-oriented data communications such as IEEE 802.11 and HIPERLAN specifications; and (3) cellular systems, which allow telecommunication and limited data accesses over wide areas [3]. Observing these trends, it can be predicted that the traffic over next-generation high-speed wireless networks will be dominated by personal multimedia applications such as fairly high-speed data, video, and multimedia traffic. This generation is known as *third-generation system* (see Table 1.1). From this viewpoint,

Table 1.1
Proposed Third-Generation Standards

	Japan (ARIB)	Europe (ETSI)	USA (cdma2000)
Multiple Access Scheme	WB DS-CDMA		
Duplex Scheme	FDD and TDD		
Channel Spacing	1.25/5.0/10.0/20.0 MHz	5.0/10.0/20.0 MHz	1.25 MHz
Frame Length	10 ms		5 and 20 ms
Data Modulation— Forward	QPSK	QPSK	QPSK
Data Modulation— Reverse	BPSK	QPSK	BPSK
Data Rates Supported	8 Kbps to 4.096 Mbps		9.6–76.8 Kbps and 1036.8 and 2073.6 Kbps
Power Control	Forward and reverse		
Power Control Rate	1,600 Hz		800 Hz

early analog cell phones are labeled as *first-generation*, and similar systems featuring digital radio technologies are labeled as *second-generation* (see Figure 1.1).

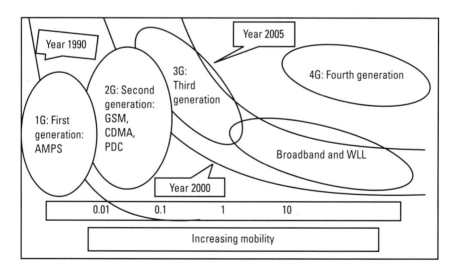

Figure 1.1 Generations of wireless networks.

The principal advantages of second-generation (digital) systems over their first-generation (analog) predecessors are greater capacity and less frequent need for battery charging [1, 2]. In other words, they accommodate more users in a given piece of spectrum and they consume less power. Second generation networks, however, retain the circuit-switching legacy of analog networks. They were all originally designed to carry voice traffic, which has little tolerance for delay jitter. Data services are more tolerant of network latencies.

The cellular network (see Figure 1.2) is an infrastructured network with wireless last hop from fixed and wired gateways. The gateways for these networks are known as base stations. A mobile terminal within these networks connects to, and communicates with, the nearest base station that is within its communication radius. As a mobile travels out of range of one base station and enters into the range of another, a handoff occurs from the old base station to the new so that the mobile is able to continue communication seamlessly throughout the network. Typical applications of this type of network include cellular systems, which allow telecommunication accesses over wide areas.

With the evolution of cellular communication, the move to digital is well underway in both the part of the spectrum used by analog wireless networks (800 MHz to 1 GHz, depending on the country) and in the newer personal communication services or personal communication network bands (in the vicinity of 2 GHz) [2]. Third-generation wireless (3G wireless) and beyond (4G mobile) have gained increased interest over the last few years. This has been fueled by a large demand for high-frequency utilization as well as a large number of users requiring simultaneous, multidimensional, high-data-rate access for applications such as mobile Internet and e-commerce. 3G wireless will use new network architecture (e.g., an all-IP network) to deliver broadband services in a more generic configuration to mobile customers. In addition, 3G

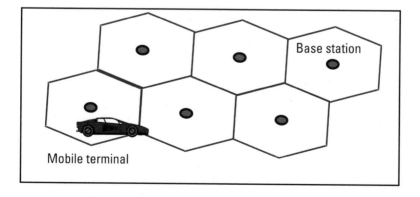

Figure 1.2 A cellular network (infrastructured network).

wireless supports multidimensional services and emerging interactive multimedia communications. Large bandwidth, guaranteed quality of service, and ease of deployment coupled with recent great advancements in semiconductor technologies for wireless applications make 3G wireless a very attractive solution for broadband service delivery. Broadband wireless, wireless mobile Internet, software radio, and reconfigurable digital radio frequency (RF) are all emerging as a result of the tremendous development in wireless semiconductors. For instance, the vision of 4G mobile is to (1) provide a technological response to accelerated growth in the demand for broadband wireless connectivity; (2) ensure seamless services provisioning across a multitude of wireless systems and networks, from private to public, and from indoor to wide area; (3) provide optimum delivery of the user's wanted service via the most appropriate network available; and (4) cope with the expected growth in Internet-based communications, new spectrum frontiers, and new market opportunities (see Table 1.2).

Table 1.2
Future Market Opportunities

Application Area	Specific Applications
Mobile office	Remote office access or database access
	File transfer
	Administrative control
	Two-way communications
	Internet browsing via the World Wide Web
Financial and retail communications	Transactions such as electronic cash or fund transfers which, generally, do not have very high communication
	Card authorization at points of sale in retail outlets
Remote control and monitoring	Traffic and transport informatics
	Traffic light monitoring and traffic movement measurements
	Route guidance systems
	Variable message signs on the roadside to inform drivers of forthcoming events or problems on the road ahead
	Trains control systems
	Vehicle fleet management
	Gas, water and electrically metering systems
	Remote monitoring and controlling of vending machines
	General telemetry systems

1.1 Mobile Wireless Networks

Wireless networks are of two types: fixed and mobile. Fixed wireless networks do not support mobility and are mostly point-to-point (e.g., microwave networks, geostationary satellite networks). On the other hand, mobile wireless networks are more versatile as they allow user mobility. Mobile wireless networks are, again, broadly classified into two distinct categories: infrastructured (cellular) and infrastructureless (ad hoc). Both aim to create a ubiquitous communication as well as computing environment where users are untethered from their information sources, that is, they get "anytime, anywhere access to information, communication, and service" with the help of the wireless mobile technologies. While cellular networks usually involve a single-hop (access only) wireless link to reach a mobile terminal, ad hoc networks normally require a multihop wireless path from a source to a destination.

The other type of mobile wireless network is the infrastructureless mobile network, commonly known as an ad hoc network (see Figure 1.3). Infrastructureless networks have no fixed gateways (routers); all nodes are capable of movement and can be connected dynamically in an arbitrary manner. Nodes of these networks function as routers, which discover and maintain routes to other nodes in the network. Example applications of ad hoc networks are emergency search-and-rescue operations, meetings or conventions in which folks wish to quickly share information, and data acquisition operations in inhospitable terrain. The comparison between these two networks is given in Table 1.3.

To add mobility support in wireless networks, the mobility management covers generally two types of mobility, namely user mobility and terminal mobility. The user mobility [4] refers to the ability of end users to originate and receive calls and access other subscribed services (telecommunication) on any terminal and on any location, and the ability of the network to identify users as they move. Personal mobility is based on the use of a unique user identity (i.e., personal number). The terminal mobility [4] is the ability of a mobile terminal to access telecommunication services from any location while in motion, and

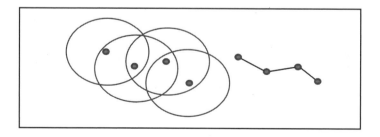

Figure 1.3 An ad hoc network (infrastructureless network).

Table 1.3
Comparison Between Infrastructured and Infrastructureless Networks

Cellular	Ad Hoc
Licensed spectrum	Unlicensed band
Standard radio signals	Adaptive signals
Network infrastructure	Ad hoc network
Symmetric two-way channel	Asymmetric information transfer
Ubiquitous coverage	Local coverage

the capability of the network to locate and identify the mobile terminal as it moves. Terminal mobility is associated with wireless access and requires the user to carry a terminal and be within the area of radio coverage.

1.2 Cellular Networks

Recent advances [1–3] in cellular communication have led to an unprecedented growth of a collection of wireless communication systems that support both personal and terminal mobility. This wide acceptance of cellular communication has led to the development of a new generation of mobile communication network, which can support a larger mobile subscriber population while providing various types of services unavailable to traditional cellular systems. Services include location independent universal phone numbering, future public land mobile telecommunications services (FPLMTS), WPBX, WLANs, telepoint phone service, and satellite communications. It is envisaged that International Mobile Telecommunications 2000 (IMT-2000) networks (previously known as FPLMTS) will evolve from the existing wireless and fixed networks by adding necessary capabilities for supporting IMT-2000 services. In a sense, IMT-2000 systems are third-generation mobile communication systems designed to provide global operation, an enhanced set of service capabilities, and significantly improved performance. While the first round of transition from analog (first generation) to digital (second generation) was designed to fix the problems (such as security, blocking, and regional incompatibilities) in the analog systems, the migration to the third generation is designed to open up a vista of entirely new services. In this generation, it is estimated that the introduction of different types of services and the establishment of new service providers will result in an unprecedented growth in the number of mobile subscribers from 15 million currently to around 60 million by 2005.

1.2.1 Cellular Network Standards

Several wireless communications systems have achieved rapid growth due to heavy market demand. Obvious examples [2] include high-tier digital cellular systems like Global System for Mobile Communication (GSM), American Digital Cellular (ADC) or IS-54, Personal Digital Cellular (PDC), and Digital Communication System at 1,800 MHz (DCS1800) for widespread vehicular and pedestrian services, and low-tier cordless telecommunication systems based on Cordless Telephone 2 (CT2), Digital European Cordless Telephone (DECT), Personal Access Communications Systems (PACS), and Personal Handy Phone System (PHS) standards for residential, business, and public cordless access applications. Although the design guidelines of such systems are quite different, their individual success may suggest a potential path to achieving a complete Personal Communications Systems (PCS) vision: integration of different PCS systems, which is referred to as "heterogeneous PCS" (HPCS). A good example of the migration from second-generation mobile systems (e.g., GSM, IS-54) to the IMT-2000 vision is the evolution from the European Telecommunication Standardization Institute (ETSI)–defined GSM system to universal mobile telecommunication systems (UMTS). The UMTS system is only one of the many new third-generation systems being developed around the world, and serves as an illustration for our current discussion. UMTS cannot be developed as a completely isolated network with minimal interface and service interconnection to existing networks. Both UMTS and existing networks will need to develop along parallel, even convergent paths, if service transparency is to be achieved to any degree. This would, in the end, allow UMTS service to be supported, although at different levels of functionality, across all networks. Another important requirement for seamless operation of the two standards is GSM-UMTS handover in both directions.

UMTS wideband code division multiple access (WCDMA) is one of the major new third-generation mobile communication systems being developed within the IMT-2000 framework. It represents a substantial advance over existing mobile communications systems. Additionally, it is being designed with flexibility for users, network operators, and service developers in mind and embodies many new and different concepts and technologies. UMTS services are based on standardized service capabilities, which are common throughout all UMTS users and radio environments. This means that personal users will experience a consistent set of services even when they roam from their home network to other UMTS operators—a virtual home environment (VHE). Users will always feel that they are connected to their home network, even when roaming. VHEs will ensure the delivery of the service provider's total environment (e.g., a corporate user's virtual work environment), independent of the user's location or the mode of access. The ultimate goal is transparency (i.e., that all networks, signaling, connections, registrations, and any other technologies should be invisible to the user), ensuring that mobile multimedia services are simple, user-friendly, and effective.

1.2.2 Cellular Architecture

The architecture of a basic cellular network is shown in Figure 1.4. The entire service space is divided into cells, where each cell is served by a base station (BS). Each BS is responsible for communicating with mobile hosts (end users) within its cell. When a mobile host changes cells while communicating, handoff occurs and the mobile host starts communicating via a new BS [5].

Each BS is connected to a mobile switching center (MSC) through fixed links. Each MSC is connected to other MSCs and the public switched telephone network (PSTN). Each MSC handles two major tasks: switching a mobile user from one base station to another and locating the current cell of a mobile user. The MSCs communicate with location registration databases such as the home location register (HLR) and the visitor location register (VLR) to provide roaming management. The HLR at each MSC is a database recording the current

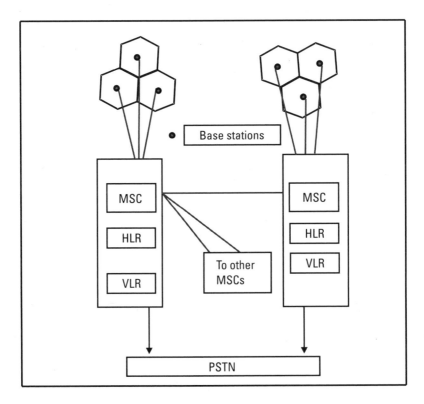

Figure 1.4 The architecture of a basic cellular network.

location of each mobile that belongs to the MSC. And the VLR at each MSC is a database recording the cell of visiting mobiles.

The distinguishing feature of cellular systems compared to previous mobile radio systems is the use of many BSs with relatively small coverage radii (on the order of 10 km or less versus 50 to 100 km for earlier mobile systems). Multiple BSs, which are a few cells apart (e.g., 5 cells, 7 cells), use the same set of frequencies simultaneously. This frequency reuse allows a much higher subscriber density per megahertz of spectrum than earlier noncellular systems. System capacity can be further increased by reducing the cell size (the coverage area of a single BS) down to an area with a radius as small as 0.5 km. In addition to supporting much higher subscriber densities than previous systems, this approach makes possible the use of small, battery-powered portable handsets with lower RF transmit power than the large, vehicular mobile units used in earlier systems. In cellular systems, continuous coverage is achieved by executing a handoff as the mobile unit crosses cell boundaries. This requires the mobile to change frequencies under control of the cellular network.

1.2.3 Medium Access

The development of low-rate digital speech coding techniques and the continuous increase in the device density of integrated circuits (i.e., transistors per unit area), have made completely digital second-generation systems viable. Second generation cellular systems based on digital transmissions are currently being used. Digitization allows the use of time division multiple access (TDMA) and code division multiple access (CDMA) as alternatives to frequency division multiple access (FDMA). With TDMA, the usage of each radio channel is partitioned into multiple timeslots, and each user is assigned a specific frequency and timeslot combination. Thus, only a single mobile in a given cell is using a given frequency at any particular time. With CDMA, multiple mobiles in a given cell use a frequency channel simultaneously, and the signals are distinguished by spreading them with different codes. One obvious advantage of both TDMA and CDMA is the sharing of radio hardware in the BS among multiple users. Digital systems can support more users per BS per megahertz of spectrum, allowing wireless system operators to provide service in high-density areas more economically. The use of TDMA or CDMA digital architectures also offers additional advantages, including the following:

- A more natural integration with the evolving digital wireline network;
- Flexibility for mixed voice and data communication, and the support of new services;
- A potential for further capacity increases as reduced-rate speech coders are introduced;

- Reduced RF transmit power (increasing battery life in handsets);
- Encryption for communication privacy;
- Reduced system complexity (e.g., mobile-assisted handoffs, fewer radio transceivers).

1.3   Ad Hoc Wireless Networks

Most of the wireless mobile computing applications today require single hop wireless connectivity to the wired network. This is the traditional cellular network model, which supports the current mobile computing needs by installing BSs and access points. In such networks, communications between two mobile hosts completely rely on the wired backbone and the fixed BSs. A mobile host is only one hop away from a BS.

At times, however, no wired backbone infrastructure may be available for use by a group of mobile hosts. Also, there might be situations in which setting up fixed access points is not a viable solution due to cost, convenience, and performance considerations. Still, the group of mobile users may need to communicate with each other and share information between them. In such situations, an ad hoc network can be formed. An ad hoc network is a temporary network, operating without the aid of any established infrastructure of centralized administration or standard support services regularly available on the wide area network to which the hosts may normally be connected [6]. Applications of ad hoc networks include military tactical communication, emergency relief operations, and commercial and educational use in, for example, remote areas or meetings where the networking is mission-oriented or community-based.

Ad hoc networks [6, 7] are envisioned as infrastructureless networks where each node is a mobile router equipped with a wireless transceiver. A message transfer in an ad hoc network environment would take place either between two nodes that are within the transmission range of each other or between nodes that are indirectly connected via multiple hops through some other intermediate nodes. This is shown in Figure 1.5. Node C and node F are outside the wireless transmission range of each other but are still able to communicate via the intermediate node D in multiple hops.

There has been a growing interest in ad hoc networks in recent years [8, 9]. The basic assumption in an ad hoc network is that two nodes willing to communicate may be outside the wireless transmission range of each other, but they are still able to communicate if other nodes in the network are willing and capable of forwarding packets from them. The successful operation of an ad hoc network will be interrupted, however, if an intermediate node, participating in a communication between two nodes, either moves out of range suddenly or switches itself off in between message transfers. The situation is worse if there is

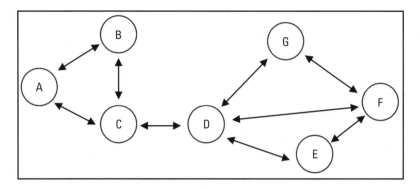

Figure 1.5 Basic structure of an ad hoc network.

no alternative path available between those two nodes. Thus, the dynamics of these networks, as a consequence of mobility and disconnection of mobile hosts, pose a number of problems in designing schemes for effective message communication between any source and destination [10].

1.4 Location Management

The growth of mobility aspects in cellular networks occurs at three different levels [3]. First, there is the spatial level, that is, users desire to roam with a mobile terminal. Second, growth occurs from the penetration rate of mobile radio access lines. Third, the traffic generated by each wireless user is constantly growing. On one hand, tetherless (e.g., cellular, ad hoc) subscribers use their mobile terminals; on the other hand, more capacity-greedy services (e.g., Internet accesses, multimedia services) arrive one after another. From these considerations, the generalized mobility features will have serious impacts on the wireless telecommunications networks. Mobility can be categorized into two areas:

- Radio mobility, which mainly consists of the handover process;
- Network mobility, which mainly consists of location management (location updating and paging).

This book will concentrate on the network mobility only.

1.4.1 Location Updating and Paging

The main task of location management [11, 12] is to keep track of the user's current location, so that an incoming message (call) can be routed to his or her mobile station (MS). Location management schemes are essentially based on users' mobility and incoming call rate characteristics. The network mobility process has to face strong antagonism between its two basic procedures: (1) updating (or registration), the process by which a mobile endpoint initiates a change in the location database according to its new location; and (2) finding (or paging), the process by which the network initiates a query for an endpoint's location (which may also result in an update to the location database). The location updating procedure allows the system to keep the user's location knowledge, more or less accurately, in order to be able to find him or her, in case of an incoming call, for example. Location updating is also used to bring the user's service profile near its location and allows the network to rapidly provide the user with his or her services. The paging process achieved by the system consists of sending paging messages in all cells where the mobile terminal could be located.

Most location management techniques use a combination of updating and finding in an effort to select the best trade-off between update overhead and delay incurred in finding. Specifically, updates are not usually sent every time an endpoint enters a new cell, but rather are sent according to a predefined strategy such that the finding operation can be restricted to a specific area. There is also a trade-off, analyzed formally, between the update and paging costs. For this purpose, the MS frequently sends location update messages to its current MSC. If the MS seldom sends updates, its location (e.g., its current cell) is not known exactly and paging is necessary for each downlink packet, resulting in a significant delivery delay. On the other hand, if location updates happen very often, the MS's location is well known to the network, and the data packets can be delivered without any additional paging delay. Quite a lot of uplink radio capacity and battery power, however, is consumed for mobility management in this case. Thus, a good location management strategy must be a compromise between these two extreme methods.

1.4.2 Mobility Models

Three mobility models, namely, the fluid flow model, the random-walk model, and the gravity model, are addressed [13]. The fluid flow model considers traffic flow as the flow of a fluid, modeling macroscopic movement behavior. The random-walk model (also known as Markovian model) describes individual movement behavior in any cellular network. The gravity model has also been used to model human movement behavior. It is also applied to regions of varying sizes, from city mobility models to national and international mobility models. Mobility traces indicate current movement behavior of users and are more realistic than mobility models. However, mobility traces for large population sizes and large geographical areas have been categorized into a hierarchy by three different scales: Metropolitan Mobility Model, National Mobility Model, and International Mobility Model.

1.4.3 Location Tracking

In a cellular network, location-tracking mechanisms may be perceived as updating and querying a distributed database (the location database) of endpoint identifier-to-address mappings [12]. In this context, location tracking has two components: (1) determining when and how a change in a location database entry should be initiated, and (2) organizing and maintaining the location database. In cellular networks, endpoint mobility within a cell is transparent to the network, and hence location tracking is only required when an endpoint moves from one cell to another. The location-tracking methods are broadly classified into two groups. The first group includes all methods based on algorithms and network architecture, mainly on the processing capabilities of the system. The second group contains the methods based on learning processes, which require the collection of statistics on subscribers' mobility behavior, for instance. This type of method emphasizes the information capabilities of the network.

1.4.4 Radio Resource Management

The problem of radio resource management is one important issue for good network performance. The radio resource management problem depends on the three key allocation decisions that are concerned with waveforms (channels), access ports (or base stations), and with the transmitter powers. Both channel derivation and allocation methods will influence the performance. The use of TDMA and CDMA are alternatives to FDMA used in the first-generation systems. With TDMA, the usage of each radio channel is partitioned into multiple timeslots, and each user is assigned a specific frequency and timeslot combination. Thus, only a single mobile in a given cell is using a given frequency at any particular time. With CDMA (which uses direct sequence spreading), multiple mobiles in a given cell use a frequency channel simultaneously, and the signals are distinguished by spreading them with different codes. The channel allocation is an essential feature in cellular networks and impacts the network performance.

1.5 Wireless Routing Techniques

A network must retain information about the locations of endpoints in the network, in order to route traffic to the correct destinations. Location tracking (also referred to as mobility tracking or mobility management) is the set of mechanisms by which location information is updated in response to endpoint mobility. In location tracking, it is important to differentiate between the identifier of an endpoint (i.e., what the endpoint is called) and its address (i.e., where the endpoint is located). Mechanisms for location tracking provide a time varying mapping between the identifier and the address of each endpoint [12].

In any communication network, procedures for route selection and traffic forwarding require accurate information about the current state of the network (e.g., node interconnectivity, link quality, traffic rate, endpoint locations) in order to direct traffic along paths that are consistent with the requirements of the session and the service restrictions of the network. Traffic sessions in wireline networks usually employ the same route throughout the session, and the route is calculated once for each session (normally, prior to the beginning of the session). Traffic sessions in mobile wireless networks, however, may require frequent rerouting because of network and session state changes. The degree of dynamism in route selection depends on several factors, such as (1) the type and frequency of changes in network and session state; (2) the limitations on response delay imposed in assembling, propagating, and acting upon this state information; (3) the amount of network resources available for these functions; and (4) the expected performance degradation resulting from a mismatch between selected routes and the actual network and session state. For instance, if the interval of time between successive state changes is shorter than the minimum possible response delay of the routing system, better performance may actually be achieved by not attempting to reroute for every state change [12]. Moreover, the routing system can decrease its sensitivity to small state changes while continuing to select feasible routes, by capturing statistical characterizations of the session and network state and by selecting routes according to these characterizations. If a state change is large enough to significantly affect the quality of service provided along the route for a session, the routing system attempts to adapt its route to account for this change, in order to minimize the degradation in service to that session.

As in stationary networks, the types of route selection and forwarding procedures employed in mobile networks depend partially upon whether the underlying switching technology is circuit-based or packet-based, and in part on whether the switches themselves are stationary or mobile. In most cellular networks, routes are computed by an off-line procedure, and calls are forwarded along circuits set up along these routes. Handoff procedures enable a call to continue when a mobile endpoint moves from cell to cell. In most mobile ad hoc networks, the mobile hosts themselves compute routes, and traffic is forwarded hop-by-hop at each switch along the route. The mobile hosts individually adjust routes according to perceived changes in network topology resulting from host movement.

In mobile networks with stationary infrastructure (i.e., cellular networks), the main component of route selection for mobile endpoints is handoff. In mobile networks with mobile infrastructure (i.e., mobile ad hoc networks), the hosts not only need to keep track of the locations of other mobile endpoints but also need to keep track of each other's location and interconnectivity as they move. Route selection requires information about the interconnectivity and services provided by the hosts as well as information about the service requirements for the session and the locations of the session endpoints. This is a difficult task, however, in such a highly dynamic environment, since the topology update information needs to be propagated frequently throughout the network. In an ad hoc network, where network topology changes frequently and where transmission and channel capacity is scarce, the procedures for distributing routing information and selecting routes must be designed to consume a minimum amount of network resources and must be able to quickly adapt to changes in network topology [12].

In cellular wireless networks, there are a number of centralized entities to perform the function of coordination and control. In ad hoc networks, since there is no preexisting infrastructure, these centralized entities do not exist. Thus, lack of these entities in the ad hoc networks requires distributed algorithms to perform equivalent functions. Designing a proper medium access control and routing scheme in this context is a challenging task which will be discussed in detail in subsequent chapters.

References

[1]   Cox, Donald C., "Wireless Personal Communications: What is it?" *IEEE Personal Communication Magazine*, Apr. 1995, pp. 20–35.

[2]   Padgett, Jay E., Gunther G. Christoph, and Takashi Hattori, "Overview of Wireless Personal Communications," *IEEE Communication Magazine*, Jan. 1995, pp. 28–41.

[3]   Tabanne, S., "Location Management Methods for Third-Generation Mobile Systems," *IEEE Communication Magazine*, Aug. 1997, pp. 72–84.

[4]   Pandya, R., "Emerging Mobile and Personal Communication System," *IEEE Communication Magazine*, June 1995, pp. 44–52.

[5]   Lin, Yi-Bing, and I. Chalmtac, "Heterogeneous Personal Communications Services: Integration of PCS Systems," *IEEE Communication Magazine*, Sept. 1996.

[6]   Johnson, D., "Routing in Ad Hoc Networks of Mobile Hosts," *Proc. IEEE Workshop on Mobile Comp. Systems and Appls.*, Dec. 1994.

[7]   Corson, S., J. Macker, and S. Batsell, "Architectural Considerations for Mobile Mesh Networking," *Internet Draft RFC Version 2*, May 1996.

[8]   Royer, E. M., and C. K. Toh, "A Review of Current Routing Protocols for Ad Hoc Wireless Networks," *IEEE Personal Communication Magazine*, Apr. 1999, pp. 46–55.

[9]   Lee, S. J., M. Gerla, and C. K. Toh, "A Simulation Study of Table-Driven and On-Demand Routing Protocols for Mobile Ad Hoc Networks," *IEEE Network Magazine*, Vol. 13, No. 4, July 1999, pp. 48–54.

[10]   Haas, Z. J., and S. Tabrizi, "On Some Challenges and Design Choices in Ad Hoc Communications," *IEEE MILCOM*, Bedford, MA, Oct. 18–21, 1998.

[11]   Akyildiz, Ian F., and Joseph S. M. Ho, "On Location Management for Personal Communications Networks," *IEEE Communication Magazine,* Sept. 1996.

[12]   Ramanathan, S., and M. Steenstrup, "A Survey of Routing Techniques for Mobile Communication Networks," *ACM/Baltzer Mobile Networks and Applications,* 1996, pp. 89–104.

[13]   Lam, Derek, Donald C. Cox, and Jennifer Widom, "Teletraffic Modeling for Personal Communications Services," *IEEE Communication Magazine*, Feb. 1995, pp. 79–87.

2

Mobility Issues

2.1 Introduction

The 3G and 4G wireless cellular systems offer a plethora of services, for example, voice, low- and high-bit-rate data, and video to mobile users (MUs) via a range of mobile terminals, operating in both public and private environments such as office areas, residences, and transportation media, independent of time, locations, and mobility patterns. To cope with the envisaged overwhelming traffic demands and to provide different services, a layered cell architecture consisting of macrocells, microcells, and picocells has been adopted in 3G wireless networks. Compared to second-generation systems and apart from the increased traffic demands, the employment of location management and handover procedures in a microcellular environment, in conjunction with the huge number of MUs, will generate a considerable mobility-related signaling load. The increase of mobility-related signaling, apart from the radio link, will have a major impact on the number of database transactions, thus causing the database to be a possible bottleneck at the fixed network side. Consequently, given the scarcity of radio resources, methods for signaling load reduction are emerging for 3G and 4G wireless networks. The analyses of different aspects of mobile wireless networks related to location management (e.g., location area planning, paging strategies), radio resource management (multiple access techniques, channelallocation schemes), and propagation (fading, handover decisions) involve mobility modeling. The accuracy of the mobility models involved in the planning of the wireless network is desirable, since it may affect the ratio of system capacity versus network implementation cost.

Three basic types of mobility models that are appropriate for the full range of the 3G and 4G wireless network design issues (e.g., location and paging area planning, handover strategies, channel assignment schemes) are introduced. The traffic models are based on call traffic data, airplane passenger traffic data, and personal transportation surveys and take into account callee distributions. Using techniques and results from transportation research, three mobility traces, to characterize movements on different scales, are also addressed: within a metropolitan area, within a national area, and at the international level. The fluid flow model, random walk model, diffusion model, and gravity model are used to model human movements in different scales in the 1G and 2G wireless networks.

2.2 Mobility Models

Teletraffic models are an invaluable tool for network planning and design [1]. They are useful in areas like network architecture comparisons, network resource allocations, and performance evaluation of protocols. Traditional traffic models have been developed for wireline networks. These models predict aggregate traffic going through telephone switches. As such, they do not include subscriber mobility or callee distributions and therefore need modifications to be applicable for modeling mobile wireless network traffic. Mobility models are required to describe movement behavior on different scales. As a general model for cellular traffic does not yet exist, most researchers resort to adding their own ad hoc mobility models to the traditional wireline models. These ad hoc mobility models seldom reflect actual movement patterns. Mobility models are required to describe movement behavior on different scales.

There are a few models for delineating the mobility of MUs. The common approaches for modeling human movements are described below. Among these are fluid flow model, diffusion model, gravity model, and Markovian model.

2.2.1 Fluid Flow Model

The fluid flow model [2, 3] conceptualizes traffic flow as the flow of a fluid. It is used to model macroscopic movement behavior. In its simplest form, the model formulates the amount of traffic flowing out of a region to be proportional to the population density within the region, the average velocity, and the length of the region boundary. This fluid model is accurate for a symmetric grid of streets and gives the crossings in only one direction across the perimeter of an area. For a region with a population density of ρ, an average velocity v of mobile terminal, and region diameter or region perimeter L, the average number of site crossings per unit time N is $N = \rho \pi L v$ for a circular cell region or $N = \rho L v / \pi$ for a rectangular cell region. The total number of crossings in and out of the area is twice this. A more sophisticated fluid model has also been formulated.

This fluid flow model considers [3] a oneway highway (semi-infinite) street that can be regarded as the location space of the interval $[0, \infty)$. There are two types of vehicles, calling and noncalling, running on the street. The vehicles of these two categories at location x and time t move forward on the highway according to a deterministic velocity $v(x, t)$, and the flow of vehicles is ensured in a single direction using assumptions $v(x, t) \geq 0$ for all x and t with $x \geq 0$ and $t \in [-\infty, +\infty]$. Without loss of generality, it is assumed that both calling and noncalling vehicles can enter and leave the highway at an y location. Two types of models have been discussed here. One of the models captures both time-dependent behavior (i.e., nonhomogeneous arrivals of vehicles) and vehicle movement on the highway. The second type captures only the spatial dynamics of the movement of the vehicles in the highway, that is, the time-homogenous fluid model instead of the nonhomogeneous time model.

2.2.1.1 Time-Nonhomogeneous Deterministic Fluid Model

Several notations have been introduced [3]: $N(x, t)$ and $Q(x, t)$ are the number of noncalling and calling vehicles in location $(0, x]$, respectively. As the model treats vehicles as a continuous fluid, $N(x,t)$ and $Q(x,t)$ are any nonnegative real numbers. In addition, $n(x,t)$ and $q(x, t)$ are the noncalling and calling density at location x and time t, respectively. That is, $n(x, t) \equiv \partial N(x, t)/\partial x$ and $q(x, t) \equiv \partial Q(x, t)/\partial x$. Furthermore, the numbers of noncalling vehicles $C^{+}_{n}(x,t)$ and $C^{-}_{n}(x,t)$, and the numbers of calling vehicles $C^{+}_{q}(x,t)$ and $C^{-}_{q}(x,t)$ are the numbers entering or leaving in location $(0, x]$ in time $(-\infty, t]$, respectively. A noncalling (calling) vehicle may enter the system, if either: (1) it is an actual arrival of a noncalling (calling) vehicle to the highway, or (2) it was a calling (noncalling) vehicle existing on the highway but with its call just terminated (started). Again, a noncalling (calling) vehicle leaves if it departs from the highway or becomes a calling (noncalling) vehicle by initiating (terminating) a call. Finally, the rate densities are $c^{+}_{n}(x,t) \equiv \partial^{2} C^{+}_{n}(x,t)/\partial x \partial t$ and $c - n(x,t) \equiv \partial^{2} C^{-}_{n}(x,t) / \partial x \partial t; c^{+}_{q}(x,t) \equiv \partial^{2} C^{+}_{q}(x,t) / \partial x \partial t$ and $c^{-}_{q}(x,t) \equiv \partial^{2} C^{-}_{q}(x,t) / \partial x \partial t$.

The evolution of noncalling and calling vehicles on the highway is governed by the partial differential equations (PDEs):

$$\partial n(x,t) / \partial t + \partial / \partial x[n(x,t)v(x,t)] = c^{+}_{n}(x,t) + c^{-}_{n}(x,t) \qquad (2.1)$$

$$\partial q(x,t) / \partial t + \partial / \partial[q(x,t)v(x,t)] = c^{+}_{q}(x,t) + c^{-}_{q}(x,t) \qquad (2.2)$$

The additional notations are used to show how these, (2.1) and (2.2), are coupled owing to calling activity. The numbers of noncalling vehicles $E^+_n(x,t)$ and $E^-_n(x,t)$, and the numbers of calling vehicles $E^+_q(x,t)$ and $E^-_q(x,t)$ are entering or leaving from the highway in location $(0, x]$ in time $(-\infty, t]$, respectively. The associated rate densities are: $e^+_n(x,t) \equiv \partial^2 E^+_n(x,t) / \partial x \partial t$ and $e^-_n(x,t) \equiv \partial^2 E^-_n(x,t) / \partial x \partial t$; $e^+_q(x,t) \equiv \partial^2 E^+_q(x,t) / \partial x \partial t$ and $e^-_q(x,t) \equiv \partial^2 E^-_q(x,t) / \partial x \partial t$.

Furthermore, $\beta(x,t)n(x,t)$ and $\gamma(x,t)q(x,t)$ are the rates at which noncalling and calling vehicles actually depart from the highway at location x at time t, respectively. Additionally, let $\lambda(x,t)n(x,t)$ be the call-initiation rate of noncalling vehicles and $\mu(x,t)q(x,t)$ be the call-termination rate of calling vehicles at location x at time t. In the stochastic model, these are stochastic intensities for individual vehicles; these are actual deterministic flow rates. The rate densities $c^+_n(x,t), c^-_n(x,t), c^+_q(x,t)$ and $c^-_q(x,t)$ are expressed in terms of these parameters. The four rate densities are

$$c^+_n(x,t) = e^+_n(x,t) + \mu(x,t)q(x,t) \tag{2.3}$$

$$c^-_n(x,t) = \beta(x,t)n(x,t) + \lambda(x,t)n(x,t) \tag{2.4}$$

$$c^+_q(x,t) = e^+_q(x,t) + \lambda(x,t)n(x,t) \tag{2.5}$$

$$c^-_q(x,t) = r(x,t)q(x,t) + \mu(x,t)q(x,t) \tag{2.6}$$

Combining these six above equations, the following coupled PDEs characterize the densities $n(x, t)$ and $q(x, t)$ in this model and can be regarded as the deterministic fluid model. The densities of noncalling and calling vehicles, $n(x, t)$ and $q(x, t)$ satisfy the coupled PDEs:

$$\partial n(x,t) / \partial t + \partial / \partial x[n(x,t)v(x,t)] = e^+_n(x,t) + \mu(x,t)q(x,t)$$
$$- [\beta(x,t) + \lambda(x,t)]n(x,t) \tag{2.7}$$

$$\partial q(x,t) / \partial t + \partial / \partial x[q(x,t)v(x,t)] = e^+_q(x,t) + \lambda(x,t)n(x,t)$$
$$- [\gamma(x,t) + \mu(x,t)]q(x,t) \tag{2.8}$$

The highway is considered to be divided into cells, labeled by i = 1, 2, 3,.... For i (1, let the boundary between cell $i - 1$ and cell i be located at $x_i - 1$ and $x_0 \equiv 0$. Furthermore, $\{y_i : i - 1,2,3...\}$ is the location of the ith entrance or exit on the highway.

This subsection is concluded by commenting on the rate densities of vehicles entering and leaving the highway for the case where vehicles can enter or leave only at entrances and exits at fixed locations, as in real vehicles. Furthermore, $\xi^i_n(t)$ and $\xi^i_q(t)$ denote the external arrival rate of noncalling and calling vehicles at the ith entrance at time t, respectively. Then,

$$e^+_n(x,t) = \Sigma_i \xi^i_n(t)\delta(x - y_i) \tag{2.9}$$

$$e^+_q(x,t) = \Sigma_i \xi^i_q(t)\delta(x - y_i) \tag{2.10}$$

where $\lim_{\varepsilon \to 0} \int_{x-\varepsilon}^{x+\varepsilon} \delta(y)dy = 1$ if $x = 0$ and 0 otherwise.

As vehicles leave the highway, $p^i_n(t)$ and $p^i_q(t)$ denote the fraction of noncalling and calling vehicles departing when they pass by the ith exit at time t, respectively. If these departing vehicles leave at the same velocity as they move forward along the highway, then

$$\beta(x,t) = v(x,t)\Sigma_i p^i_n(t)\delta(x - y_i) \tag{2.11}$$

$$\gamma(x,t) = v(x,t)\Sigma_i p^i_q(t)\delta(x - y_i) \tag{2.12}$$

2.2.1.2 Time-Homogeneous Deterministic Fluid Model

This time-homogeneous deterministic model [3] considers the system that has reached a steady state with respect to time. As a result, all system variables and parameters become independent of time. A stronger proportionality result for a time-dependent setting can be set out to determine the proportion of vehicles arriving to the highway to those exiting from the highway. For some $x_0 \geq 0$,

$$\text{if } \lambda(x) = \lambda, \mu(x) = \mu, \beta(x) = \gamma(x), \text{ and } e^+_q(x) / e^+_n(x) = \lambda / \mu, \tag{2.13}$$

for all $x \geq x_0$, and

$$\text{if } q(x_0) \text{ is finite and } q(x_0)/n(x_0) = \lambda / \mu \tag{2.14}$$
then

$$q(x)/n(x) = \lambda / \mu \text{ for all } x \geq x_0 \tag{2.15}$$

The above set of expressions carry a clear physical meaning and are natural for the time-homogeneous model. When $\lambda(x) = \lambda$ and $\mu(x) = \mu$, vehicles initiate and terminate calls at rates independent of their locations. The condition $\beta(x) = \gamma(x)$ indicates that a vehicle departs from the highway at the same rate, regardless of whether it is a calling or a noncalling vehicle. The ratio

$e^{+}_{q}(x) / e^{+}_{n}(x) = q(x_{0}) / n(x_{0}) = \lambda / \mu$ means that the proportion of vehicles arriving to the highway at location x which are calling vehicles is identical to that of existing vehicles at location x_{0}, which, in turn, is equal to the ratio λ / μ.

One of the limitations of the fluid model is that it describes aggregate traffic and therefore is hard to apply to situations where individual movement patterns are desired, (e.g., when evaluating network protocols or data management schemes with caching). Another limitation comes from the fact that since average population density and average velocity are used, this model is more accurate for regions containing a large population.

2.2.2 Diffusion Model

A more sophisticated fluid model has been formulated by characterizing the flow of traffic as a diffusion process [4]. A time-varying location probability distribution in conjunction with a Poisson page-arrival model is used to formulate the paging and registration model in terms of a set of timeout parameters τ_{m}. Each timeout parameter is defined as the maximum amount of time to wait before registering given in the last known location was m. The simple case of memoryless motion is chosen for the clarity of the model. To illustrate this method on a time-varying Gaussian user location arises as a result of isotropic random user motion. A number of motion models which are specified in terms of independent increments result in Gaussian distributions on location probability. This model would, for example, be used to obtain the minimum average paging; each location would be searched in the decreasing order of probability. The mean distribution with equally symmetric locations to either side of the mean would give the most likely location.

2.2.3 Gravity Model

Gravity models [1] have been used to model human movement behavior and applied to regions of varying size, from city models to national and international models. In its simplest form, the amount of traffic $T_{i,j}$ moving from region i to region j is described by: $T_{i,j} = K_{i,j} P_{i} P_{j}$ where P_{i} is the population in region i, and $\{K_{i,j}\}$ are parameters that have to be calculated for all possible region pairs (i, j). The different variations of this model usually have to do with the functional form of $K_{i,j}$. For example, analogous to Newton's gravitational law, $K_{i,j}$ can be specified to have inverse square dependence with the distance between zones i and j.

In the above expression, the model describes aggregate traffic and therefore suffers from some of the same limitations as the fluid model. If P_{i} is interpreted as the attractivity of region i, however, and $T_{i,j}$ as the probability of movements between i and j, then the model describes individual movement behavior. Using this approach, the parameters $\{T_{i,j}\}$ also have to be calculated from the traffic data in addition to $\{K_{i,j}\}$. The advantage of the gravity model is that frequently visited locations can be modeled easily since they are simply regions with large attractivity. The main difficulty with applying the gravity model is that many parameters have to be calculated, therefore it is hard to model geography with many regions.

2.2.4 Random Walk Model

The random walk model [5] is explained with the help of a cellular wireless radio system with N cells. Two cells are called neighboring cells if a mobile user can move from one of them to the other without crossing another cell. A ring cellular topology consists of cells where cells i and $i + 1$ are neighboring cells. Thus, a mobile user that is in cell i, can only move to cells $i + 1$ or $i - 1$ or remain in cell i. To model the movement of the mobile users in the system, we assume that time is slotted, and that a user can make at most one move during a slot. The movements will be assumed to be stochastic and independent from one user to another. Three update strategies are taken: (1) time-based update, (2) movement-based update, and (3) distance-based update to understand the Markovian random walk model in those strategies. In the Markovian model, during each slot, a user can be in one of the following three states: (1) the stationary state S, (2) the right-move state R, or (3) the left-move state L. Assume that a user is in cell i at the beginning of a slot. The movement of the user during that slot depends on the state as follows. If the user is in state S then it remains in cell i, if the user is in state R then it moves to cell $i + 1$, and if the user is in state L then it moves to cell $i - 1$. Let $X(t)$ be the state during slot t. Assume that $\{X(t); t = 0; 1; 2;...\}$ is a Markov chain with transition probabilities $p_{k,l} = \text{Prob}[X(t+1) = l = X(t) = k]$ as follows: $p_{R,R} = p_{L,L} = q$, $p_{L,R} = p_{R,L} = v$, $p_{S,R} = p_{S,L} = p$, $p_{L,S} = p_{R,S} = 1 - q - v$ and $p_{S,S} = 1 - 2p$ (see Figure 2.1).

In time-based update, each user transmits an update message every T slots, while in movement-based update, each mobile user transmits an update message whenever it completes M movements between cells, and finally in distance-based update, each user transmits an update message whenever the distance, in terms of cells, between its current cell and the cell in which it last reported is D. The act of a user sending an update message is referred to as reporting.

For simplicity, only the random walk model for movement-based update is addressed in this section [5].

Let $Y(t)$ be the distance between the cell in which the user is located in slot t and the cell in which the user last transmitted an update message. As before, positive (negative) $Y(t)$ indicates that the user is to the right (left) of the cell from which an update message was last transmitted. Clearly, the interval is $-(M-1) \le Y(t) \le M-1$. Let $L(t) = \max \{\pi \le t|$ The user reported in slot $\pi\}$. Let $I(t)$ be the

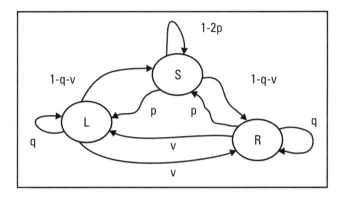

Figure 2.1   A state diagram of the Markov walk model.

number of movements that the user has made during the slots $L(t)$; $L(t) + 1,\ldots,$ $t-1$ (if $t - 1 < L(t)$, then $I(t) = 0$). Clearly, $0 \leq I(t) \leq M - 1$. To compute the expected number of update messages per slot transmitted by the user U_M, we focus on the Markov chain $\{(I(t), X(t)), t = 0, 1, 2,\ldots\}$ with the stationary probabilities $Q_{m,x} = \lim_{t \to \infty} Prob[I(t) = m, X(t) = x]$; $m = 0, 1,\ldots, M-1, x \in \{S, R, L\}$. The following balance equations will be used shortly:

$$Q_{i-1,R} + Q_{i-1,L} = Q_{i,R} + Q_{i,L}$$
$$i = 0, 1,\ldots, M-1$$

$$(1 - q - v)(Q_{i-1,R} + Q_{i-1,L}) = 2pQ_{i,S}$$
$$i = 0, 1,\ldots, M-1$$

where $i - 1$ is computed modulu M. The user transmits an update message at slot t if and only if $I(t-1) = M-1$ and $X(t-1)$ is either R or L. Therefore,

$$UM = QM - 1, R + QM - 1, L = QM - 1, L = Q0, R + Q0, L = 1 / M$$
$$- Q0, S = 1 / M - (1 - q - v)2p * (QM - 1, R + QM - 1, L) = 1 / M$$
$$- ((1 - q - v) / 2p)UM$$

The second and fourth equalities follow from the above balance equations. The third equality follows from the fact that $Q_{m,R} + Q_{m,S} + Q_{m,L} = 1/M$ for all m, which follows from symmetry considerations. Solving for U_M, we have $U_M = 2p/M(1 + 2p - q - v)$. To compute the expected number of searches necessary to locate a user V_M, we will consider an embedded Markov chain that ignores the states in which the user does not move. To that end, let $J(t, t')$ be the number of movements that the user has made during the slots $L(t)$, $L(t) + 1,\ldots, t'-1$. Recall that t_s is the slot in which a search occurs. Let $t_m = \max \{t \geq L(t_s) \mid J(t_s, t) = m\}$. The embedded Markov chain is $\{(Y(t_m), X(t_m)), m = 0, 1, 2,...\}$. Let $P_m(d, x \mid x')$ $= Prob[Y(t_m) = d; X(t_m) = x \mid X(t_0) = x']$. From the definition of t_m, it follows that $Y(t_0) = 0)$. Define $P'_m(d, x \mid x') = P_m(d, x \mid x') + P_m(-d, x \mid x')$ *for* $d > 0$, and let $P'_m(d \mid x') = P'_m(d, R \mid x') + P'_m(d, L \mid x')$. By symmetry considerations, we have that $P'_m(d \mid R) = P'_m(d \mid L)$ for all $d > 0$ and $m \geq 0$. The probability that a user will be at an absolute distance d (from the cell from which an update message was last transmitted) at time t_m, namely after m movements, is, therefore,

$$P'_m(d) = P'm(d \mid R) \, Prob[X(t_0) = R] + P'm(d \mid L) \, Prob[X(t_0) = L] = P'm(d(R)$$

Returning to the Markov chain,

$$Prob[Y(t_s) = d] = \sum_{M=0}^{M-1} Prob[Y(t_s) = d \mid I(t_s) = m]^* \, Prob[I(t_s) = m]$$

$$= \sum_{M=0}^{M-1} P'_m(d)^* \, 1/M$$

Therefore, the expected number of searches required to locate the user is

$$S_M = 1 + 1/M \sum_{M=0}^{M-1} \sum_{d=1}^{M-1} DP_m(d)$$

where the latter sum is taken only for even $d + m$.

To complete the computation, we only need to have the quantities $P_m(d, R \mid R)$ and $P_m(d, L \mid R)$ for $0 \leq d, m \leq M-1$, and

$$P_0(0, R \mid R) = (1 + q - v)/2$$

$$P_0(0, L \mid R) = (1 - q + v)/2$$

$$P_0(d, x \mid R) = 0, \, d \neq 0; \, x \in \{R, L\}$$

$$Pm(d, R|R) = (1 + q - v)/2 * Pm - 1(d - 1, R|R)$$
$$+ (1 - q + v)/2 * Pm - 1(d + 1, L|R)$$

$$m \geq 1, \, (M-1) \leq d \leq M-1$$

$$Pm(d,L|R) = (1-q+v)/2 * Pm-1(d-1,R|R)$$
$$+(1+q-v)/2 * Pm-1(d+1,L|R)$$

$$m \geq 1, (M-1) \leq d \leq M-1$$

These probabilities can be computed recursively from the above relations. Using transform techniques, one may obtain expressions for these probabilities.

2.3 Mobility in 3G Systems

2.3.1 Metropolitan Mobility

The three basic types of modeling are appropriate for the discussion of metropolitan mobility. The three types of models city area, area zone, and street unit are introduced [6].

2.3.1.1 City Area Model

The city area model consists of a set of area zones connected via high-capacity routes. Candidate output parameters may include the user distribution per area zone versus time, the crossing rate per area zone, and the percentage of nonmoving and moving users (car passengers, pedestrians) for each area zone versus time. This model describes user mobility and traffic behavior within a city area environment. The need to analyze user mobility behavior over large-scale geographical areas is raised by location-management-related aspects. Network planning purposes impose the use of city area models representing *specific* cities (i.e., based on geographical databases, demographic data, and existing transportation studies). On the other hand, *typical* city area models are required for the evaluation of proposed system design alternatives [6]. Although each individual city area exhibits specific characteristics (e.g., population distribution, distribution of Moving Attraction Points, street network), some generic characteristics can be observed in most contemporary cities; for example:

- Cities are usually developed in such a way that densely populated areas (urban areas) surround a city center (high density of workplaces and shopping centers). While moving toward the city edges, the population density gradually decreases (suburban and rural areas).

- The street network supports two movement types: radial (i.e., from the city center toward the edge of the city and vice versa) and peripheral (see Figure 2.2).

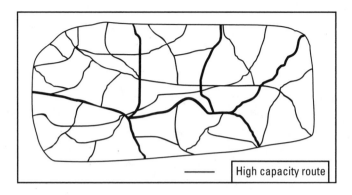

Figure 2.2 City area model.

2.3.1.2 Area Zone Model

The area zone model consists of a street network and a set of building blocks. It may be utilized for the estimation of the probability distribution function (PDF) of user residence time in an area zone or the PDF of user crossing time in an area zone, for example. Evaluating the various radio resource management schemes requires knowledge of the MU location with an accuracy of a microcell area. The model described in this section considers an area zone consisting of a set of building blocks and a street network (see Figure 2.3) covered by several microcells. Similar to the city area model, a specific area zone model can be developed for network planning purposes, while a typical area zone model can be used for research. To derive a typical area zone model, regular-shaped building blocks and a regular street-network graph can be considered. The latter leads to the well-known Manhattan grid, for example, according to which the

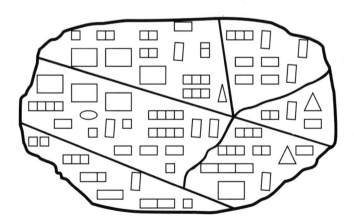

Figure 2.3 Area zone model.

square-shaped building blocks and an orthogonal grid street network represent an area zone [6].

2.3.1.3   Street Unit Model

The street unit model considers three street types: highways, streets with traffic-light-controlled flow, and high- and low-priority streets. Candidate output parameters may include the PDF of car density and car speed in a street segment, for example, or the PDF of car residence time in a street segment. This model describes the mobility behavior of (moving) MUs (pedestrians, passengers) with an accuracy of a few meters. To develop such a model, a very detailed analysis of car and pedestrian motion and street type under any vehicular traffic conditions is needed. Figure 2.4 is taken from the work of John Markoulidakis et al. [6].

2.3.2   National Mobility Model

The national mobility model [1] characterizes movement behavior between metropolitan areas in the state. Each site object now represents a metropolitan area. This model characterizes traffic volume flowing between two sites as a function of the population in each site and the distance separating them. The traffic volume in a metropolitan site is modeled using a variation of the gravity model. Instead of developing a detailed gravity model that could describe the traffic very precisely, a simplified version of the gravity model with a small

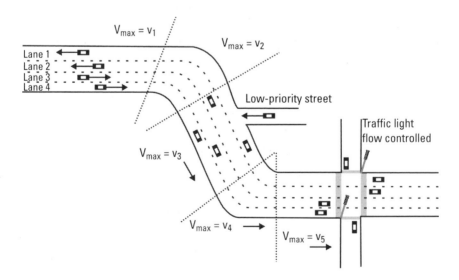

Figure 2.4   Street unit model [6].

number of parameters is used to model [1] this movement. Assuming symmetric traffic flow, (i.e., the traffic volumes between any two metropolitan areas are the same along both directions), the national mobility model is a realistic and reasonable model for movements between metropolitan areas. In addition, this model [1] is relatively insensitive to parameter variations and is reasonable to use in estimating traffic volumes in the future or for other geographies.

2.3.3   International Mobility Model

The international mobility model [1] characterizes movement behavior between one country and other countries. Each site object in this model represents a country. Compared to the national mobility model, the international gravity model is missing the inverse dependence on distance. One reason is that there is uncertainty in defining distances between countries. Defining the distance between the United States and Canada, two large territories, is a good example. In any case, the goal is to have a simple, easy-to-use, realistic model for international movement traffic.

References

[1]   Lam, Derek., Donald C. Cox, and Jennifer Widom., "Teletraffic Modeling for Personal Communications Services," *IEEE Communications Magazine*, Feb. 1997, pp. 79–87.

[2]   Frost, V. S., and B. Melamed, "Traffic Modeling for Telecommunications Networks," *IEEE Communications Magazine*, March 1994, pp. 70–81.

[3]   Leung, K. K., W. A. Massey, and W. Whitt, "Traffic Models for Wireless Communication Networks," *IEEE JSAC*, Oct. 1994, pp. 1353–1364.

[4]   Rose, C., "Minimizing the Average Cost of Paging and Registration: A Timer-Based Method," *ACM J. Wireless Networks*, Feb. 1996, pp. 109–116.

[5]   Bar-Noy, A., and I. Kessler, "Mobile Users: To Update or Not To Update?" *Proc. INFOCOM 94*, June 1994, pp. 570–576.

[6]   Markoulidakis, John G., et al., "Mobility Modeling in Third-Generation Mobile Telecommunications Systems," *IEEE Personal Communications*, Aug. 1997, pp. 41–56.

3

Radio Resource Management

The rapid increase in the size of the wireless mobile community and their demands for high-speed, multimedia communications stands in clear contrast to the limited spectrum resource that has been allocated to them in international agreements. Comparing market estimates for wireless mobile communication and considering recent proposals for wideband multimedia services with the existing spectrum allocations shows that spectrum resource management remains an important topic for the foreseeable future. Efficient spectrum resource management is, therefore, of paramount importance. In this chapter, we begin with a brief introduction to what is meant by radio resource, then present an overview of the solutions to the radio resource management (RRM) problem and finally, outline the key problems of resource management in next-generation wireless networks.

The network under consideration has already been introduced in Chapter 1. For the sake of continuity, we reiterate it briefly here. It consists of a fixed network part and a wireless access system. The fixed network provides connections between BSs, which in turn provide the wireless connections to the mobile terminals (MTs). BSs are normally distributed over the complete geographical area (service area) where the mobile users are provided with communication services. In two-way communication systems (such as mobile telephone systems), links have to be established both from the BS to the mobile (downlink or forward link) and from the mobile terminal to the BS (uplink or reverse link). The propagation situation is quite different, particularly in wide-area cellular phone systems, where the BS usually has its antennas at some elevated location, free of obstacles. The terminals, on the other hand, are usually located amidst buildings and other obstacles, creating shadowing and multipath reflections.

In the future, mobile operators will benefit from the different coverage and capacity characteristics of several air interface technologies such as WCDMA, GSM, EDGE, WLAN, and UTRAN TDD. A functional layer view of the network, with respect to various standards, (shown in Figure 3.1) clearly indicates that, since the RRM layer sits on the radio interface layer, a clear understanding of the radio propagation mechanism is an essential prerequisite for formulating the radio resource management problem. This is why we begin this chapter with an overview of the radio propagation models.

3.1 Radio Propagation

Wireless communication relies on electromagnetic waves to carry information from one point to another. This is commonly known as radio wave propagation in fixed and mobile wireless communication. The first step in the process of a new radio system design is to determine BS arrangement and a frequency plan, both of which are chiefly dependent on environmental characteristics. An accurate estimation of the propagation losses provides a good basis for a proper selection of BS locations and a proper determination of the frequency plan. By knowing propagation losses, one can efficiently determine the field signal strength, signal-to-noise ratio (SNR), carrier-to-interference ratio (CIR), for example. An accurate prediction of the field strength level is a very complex and difficult task [1]. Moreover, mobile communication is more complicated than radio systems with fixed and carefully positioned antennas. The antenna at a mobile terminal is low and has very little clearance. Thus, the transmission path between the transmitter and the receiver can vary from simple direct line of sight (LOS) to one that is severely obstructed by buildings, foliage, and the terrain.

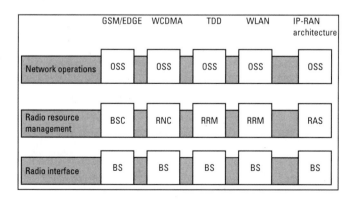

Figure 3.1 Architecture view of network functional layers.

Coverage problems due to various propagation effects put a lower limit on the number of BSs required.

The mobile radio channel is evaluated from statistical propagation models: specific terrain data is considered, and channel parameters are modeled as stochastic variables. The mean signal strength for an arbitrary transmitter-receiver (T-R) separation is useful in estimating the radio coverage of a given transmitter, whereas measures of signal variability are key determinants in system design issues, such as antenna diversity and signal coding. The modeling of the propagation path needs to take into account a number of effects. These include the following:

1. *Path loss:* The signal gets reduced in power with the distance it traverses following an inverse square law.

2. *Shadowing:* Scattering environments along various propagation paths will be different, causing variations with respect to the nominal value given by the path loss. Some paths will suffer increased loss, while others will be less obstructed and have an increased signal strength. This phenomenon is called shadowing or slow fading and is said to follow log-normal fading statistics [1].

3. *Number of multipath components and distribution of their envelopes:* These effects are caused by the local scattering environment around the MT or BS or both.

4. *Temporal fading:* The cause of this type of fading can be attributed to large-scale fading or small-scale fading. Large-scale fading or shadowing has path loss effects (explained in Section 3.1.2). Small-scale fading manifests itself in two mechanisms, namely, signal dispersion and time variant behavior of the channel.

5. *Correlation:* Multipath components generated by a single area of local scatterers may show considerable correlation, with the correlation depending heavily on the assumptions made concerning the spatial distribution of local scattering elements.

3.1.1 Path Loss

If a wireless channel's propagating characteristics are not specified, it is generally inferred that the signal attenuation versus distance behaves as if the propagation takes place over a terrain. The model of free space treats the region between the transmitting and the receiving antennas as being free of all objects that might absorb or reflect RF energy. It is further assumed that, within this region, the atmosphere behaves as a perfectly uniform and nonabsorbing medium. Furthermore, the earth is treated as being infinitely far away from the propagating signal. In this idealized free-space model, the attenuation of RF energy behaves according to an inverse-square law. Ideal free-space propagation of radio waves is governed by the following formula:

$$P_r(d) = P_t G_t G_r \xi^2 \; / \; ((4\pi)^2 \, d^2)$$

where
d = distance between the transmitter and the receiver
P_r = received power
P_t = transmitted power
G_t = transmitter antenna gain
G_r = receiver antenna gain
ξ = wavelength in meters

Spreading and interaction with propagation environment attribute to attenuation suffered by the signal. The received power expressed in terms of transmitted power is attenuated by a factor P_L. This factor is called the free-space loss or path loss.

The path loss P_L in decibels at a distance d from the transmitter in the far field region is given by the following:

$$P_L = -10[G_t G_r \xi^2 \; / \; ((4\pi)^2 \, d^2)]$$

Electromagnetic waves, while propagating, may undergo reflection, scattering, or diffraction. Reflection occurs when the wave impinges on a smooth surface with very large dimensions compared to the wavelength of the RF signal. Scattering is caused by a large rough surface or any surface whose dimensions are of the order of the wavelength or less, causing the reflected energy to spread out. In an urban environment, typical signal obstructions that yield scattering are foliage, street signs, and lampposts. The phenomenon is called diffraction when a dense body obstructs the LOS path between the transmitter and the receiver with large dimensions compared to the wavelength, causing secondary waves to be formed behind the obstructing body. It is often termed shadowing because the diffracted field can reach the receiver even when shadowed by an impenetrable obstruction (i.e., the signal can "bend" around the obstacle to some extent).

Since there is frequently no LOS between the transmitter and the receiver, the received signal is a sum of components that stem from several previously described phenomena and is quite variable with respect to time and especially with respect to the receiver or transmitter displacement. Even a displacement of just a fraction of the wavelength can cause the signal level to change by more than 30 dB. These fluctuations are known as short-term or multipath fading. On the other hand, the local average of the signal varies slowly with the displacement. These slow fluctuations depend mostly on environmental characteristics, and they are known as long-term fading.

3.1.2 Shadowing Effects

Usually shadowing effects happen when a signal suffers diffraction at some solid hindrance such as a tree or the top of a building in a city area. In an urban area, there will be many buildings blocking the LOS propagation. The simplest, albeit approximate, method of calculation of field at the point where the MT is residing is to apply the simple wedge model at the point of maxima of the curve defining the upper edges of obstacles of the wedge surface. In most applications of wireless communication, except very near the BSs, the transmission loss over a flat ground is characterized by an inverse fourth power relationship of received power with distance from the BS antenna.

Thus, wedge diffraction theory [2] provides an explanation of shadowing effects, where each building is modeled as a wedge with respect to the flat ground. It is directly related to the inverse fourth power of the distance in the flat ground case. At first, the equation of the curve defining the upper edges of the buildings is found using a least mean square fit. Then the maxima of the curve are determined. Then the wedge model of diffraction is applied at this point. A simple wedge diffraction formula [2] is normally used to find the shadowing loss with respect to the flat ground case with the MT located in a deep shadow region. With respect to flat ground, the wedge introduces two new transmitter images that are created by the wedge surface near the receiver. Diffracted signals emanate from the apex of the wedge and provide radiated signals in the shadow region where the MT is located. Figure 3.2 explains the definition of geometry for shadow boundary approximation.

3.1.3 Reciprocity

In a radio communication, propagation paths are reciprocal. Reciprocity implies that the pattern, terminal impedance, and directivity of an antenna are the same when transmitting or receiving. These properties do not vary while changing from transmitting to receiving mode or vice versa. The reciprocity theorem states that: "if a current I_1 at the terminals of antenna 1 induces a voltage V_{21} at the open terminals of antenna 2, and current I_2 at the terminals of the antenna 2 induces a voltage V_{12} at the open terminals of antenna 1, then $I_1 = I_2$ implies $V_{21} = V_{12}$." This theorem plays a role in communications between BSs and MTs. Two-way communication requires inbound (mobile-to-fixed) as well as outbound (fixed-to-mobile) communications. The inbound and outbound channels are also known as the uplink and downlink, respectively. Reciprocity refers to the relationship between uplink and downlink channels. Usually, the

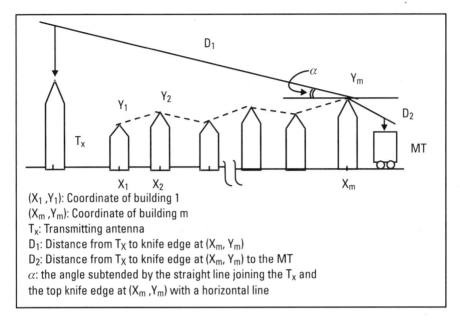

(X_1, Y_1): Coordinate of building 1
(X_m, Y_m): Coordinate of building m
T_x: Transmitting antenna
D_1: Distance from T_X to knife edge at (X_m, Y_m)
D_2: Distance from T_X to knife edge at (X_m, Y_m) to the MT
α: the angle subtended by the straight line joining the T_x and
the top knife edge at (X_m, Y_m) with a horizontal line

Figure 3.2 Geometry for shadow boundary approximation.

interference situation in the uplinks and downlinks will be different since there are many terminals, varying locations and quite a few BSs at fixed locations. That is why cellular uplink and downlink paths are normally on different frequencies. The consequence of the mobile antenna being low is that the mobile is more likely to suffer interference.

Multipath scatters mostly occur in the immediate vicinity of the mobile antenna. The BS receives, more or less, a transversal electromagnetic wave, whereas the MS receives a superposition of a set of reflected waves from random angles. Two antennas at the mobile terminal are likely to receive uncorrelated signal powers if their separation is more than a wavelength. At the BS site, however, all reflections arrive from almost identical directions. Therefore, diversity at the BS requires much larger separation of the antennas to ensure uncorrelated received signal powers at the two antennas. For the same reason, antenna directivity has different effects at the mobile terminal (MT) and the BS.

In a cellular network, shadow fading of the wanted signal received by the MS is likely to be correlated with the shadow fading of the interference caused by other BSs, or in a spread-spectrum network, with the shadowing of simultaneously transmitted signals from the same BS. In contrast to this, at the BS, shadow fading of the wanted signal is mostly statistically independent from shadow fading of the interference. However, experimental results for correlation of shadow attenuation are scarce. In full-duplex operation, multipath fading of inbound and outbound channels, which operate at widely different frequencies, may be uncorrelated. This will particularly be the case if the delay spread is large.

In a practical multiuser system with intermittent transmissions, inbound messages are sent via a multiple-access channel, whereas in an outbound channel, signals destined for different users can be multiplexed. In the latter case, the receiver in a MS can maintain carrier and bit synchronization to the continuous incoming bit stream from the BS, whereas the receiver in the BS has to acquire synchronization for each user slot. Moreover, in packet-switched data networks, the inbound channel has to accept randomly occurring transmissions by the terminals in the service area. In cellular networks with large traffic loads per BS, spread-spectrum modulation can be exploited in the downlink to combat multipath fading, whereas in the uplink, the signal powers from the various mobile subscribers may differ too much to effectively apply spread-spectrum multiple access unless sophisticated adaptive power control techniques are employed.

3.1.4 Indoor Wireless

There are several causes of signal corruption in an indoor wireless channel. The primary causes are signal attenuation due to distance, penetration losses through walls and floors, and multipath propagation. Signal attenuation over distance is observed when the mean received signal power is attenuated as a function of the distance from the transmitter (the free-space loss described above). Multipath results from the fact that the propagation channel consists of several obstacles and reflectors. Thus, the received signal arrives as an unpredictable set of reflections or direct waves or both, each with its own degree of attenuation and delay. The delay spread is a parameter commonly used to quantify multipath effects. Multipath leads to variations in the received signal strength over frequency and antenna location. For wireless LANs this could mean that an antenna placed in a local multipath null, remains in fade for a very long time. Measures such as diversity are needed to guarantee reliable communication irrespective of the position of the antenna. Wideband transmission, (e.g., direct sequence CDMA), could provide frequency diversity.

3.2 Radio Resource (Spectrum Allocation)

Clearly the radio resource (i.e., air interface) will continue to be the most limited resource in mobile operators' networks. Managing these multiple resources together, as a single resource pool, will be the key to maximizing the utilization of the operator's resources.

3.2.1   Radio Frequency Spectrum Allocation

In the United States, the Federal Communications Commission (FCC) and the National Telecommunications and Information Administration (NTIA) manage the radio frequency spectrum through a system of frequency allocations, allotments, and assignments. The entire radio spectrum is divided into blocks, or bands, of frequencies established for a particular type of service by the process of frequency allocation. Further, these general allocations can be subdivided into bands designated for a particular service, or allotment. Within these subdivided bands, specific channel plans may be implemented. For example, allocations made to the land mobile service are divided into allotments for business users, public safety users, and cellular users, with each group allotted a portion of the band in which to operate. Assignment refers to the final subdivision of the spectrum in which a party gets an assignment, or license, to operate a radio transmitter on a specific channel or group of channels at a particular location under specific conditions. The FCC also issues some licenses on a more general geographic basis. The FCC has listed U.S. allocations in a table of frequency allocations that can be found in Section 2.106 of the Commission's rules (http:// www.fcc.gov/oet/info/rules).

3.2.2   International Allocations

International coordination of spectrum is a critical component of the spectrum allocation process because radio waves transcend national borders and because of the increasing number of global services. The radio communication conferences of the International Telecommunication Union (ITU) are the principal mechanisms for international spectrum allocation. The ITU's table of allocations represents a global consensus that reflects the needs of ITU memberstates. The scheduling of conferences every two to three years with a specified agenda keeps the table of allocations flexible and current. Its general success is evidenced by, for example, the allocations provided during the 1992 conference for mobile-satellite services that are now coming into use. These conferences may address any service throughout the entire radio spectrum, depending on the agendas set by the ITU Council. In addition, there are also Regional Radio Communication Conferences, which meet as necessary and have a restricted agenda devoted to specific services for the ITU region concerned. Based on the agreements reached at these conferences, the ITU publishes the international Radio Regulations, which include allocations and technical rules for radio operation for each of the three regions of the world (http://www.fcc.gov/ ib/WRC00). The ITU further designates such allocations as primary or secondary.

- *Primary allocations* grant priority to specific services in using the allocated spectrum. When there are multiple primary services within a frequency band, they all have equal rights. A station, however, has the right to be protected from any others that start operation at a later date.

- *Secondary allocations* are made for services that must protect all primary allocations in the same band. Services operating in secondary allocations must not cause harmful interference to, and must accept interference from, primary service stations. All secondary service stations have equal rights among themselves in the same frequency band.

3.2.3 Financing for Spectrum Management

Every country faces challenges in ensuring that its spectrum regulatory body has sufficient funding to meet the costs of spectrum management. License fees represent one way to improve the economic and technical efficiency of national spectrum management. At the FCC, two types of fees are collected, application fees and regulatory fees, to cover the cost of allocating the use the spectrum. In addition to paying for the administrative cost of managing the use of the spectrum, the fees may also serve to discourage the filing of frivolous applications. If set too high, however, fees can result in underutilization of the spectrum (http://www.fcc.gov/fees).

3.2.4 Spectrum Monitoring and Enforcement

Effective spectrum monitoring and enforcement requires tools to ensure adherence to spectrum allocation and use regulations, as well as identification and elimination of interference. The United States uses a variety of tools to monitor spectrum use and enforce adherence to U.S. rules and regulations. Among these tools are (1) databases of information on licensed systems; (2) information, in the form of national rules on general licensing and technical requirements concerning specific services; (3) electronic equipment for determining the sources of interference and illegal radio operations; and (4) regulatory mechanisms for assessing penalties on licensees not complying with regulations.

In both the domestic and international context, maintaining a database of relevant technical parameters on individual communications systems is one tool necessary for monitoring compliance with spectrum use regulations. In cases of complaints of noncompliance with regulations, the regulator need merely access the database to obtain technical parameters to help determine whether a station is in compliance.

In some cases, it is not practical to maintain a database on all licensed systems and their components. For example, many systems communicate with consumer handsets that are not individually licensed. In these cases, codifying general licensing conditions and technical requirements in a rulebook provides a basis for determining compliance. With either a database of actual system parameters or a rulebook stating required conditions, spectrum regulators can conduct engineering and managerial analyses to ensure that operators comply with the relevant technical rules for spectrum use.

Identifying interference sources is also a critical component of ensuring compliance with spectrum use rules. At the FCC, the electronic direction finder has proven to be highly effective for pinpointing interference sources and taking measures to eliminate interference. If an interference complaint is received, the direction finder can be used to locate an interfering transmitter. In domestic cases of interference, a letter can be sent to the operator of the interfering station or a physical investigation of the transmitting site can be made. In the case of interference from international sources, a special note is sent to the administration of the offending operator, informing that administration of the infringement.

In cases of repeated interference or illegal operations, the FCC has the authority to assess penalties on the offending party. For this type of action to be effective, penalties related to specific violations must be clearly defined and publicized. Depending on the severity of the infraction, penalties can include warnings, fines, revocation of license, equipment seizure, and, in severe cases, incarceration.

There is an enormous pent-up demand for access to radio waves, and high barriers to entry imposed by regulators are stifling substantial economic growth. But this is because the spectrum allocation is working precisely as planned. Regulation serves to protect incumbent interests for a modest surcharge, a public interest tax that the tax collector more often than not forgets to collect.

The FCC's continuing (mis)management of the TV Band is noteworthy when the TV Band is being reduced to irrelevance. A linear extrapolation of the prices paid for PCS licenses in a January 2001 FCC auction imply that the spectrum allocated to the TV Band would fetch up to $470 billion—if they could be used for something other than TV. More importantly, consumer benefits are likely to easily exceed this number. For instance, the GSM association is attempting to gain widespread acceptance of the spectrum strategy by the following:

- Requesting administrators and lobbying regional telecommunication organizations to release the designated by the ITU in 1992;

- Inviting administrators and regional telecommunication organizations to support and propose the Association's views on and associated regulatory at World Radiocommunication Conference (WRC-2000).

Concerning the first item above, regulatory authorities are encouraged to release sufficient IMT-2000 core-band spectrum to facilitate the introduction of 3GSM IMT-2000 services utilizing the GSM platform from 2002. A good starting point would be 2×15 MHz + 5 MHz (i.e., 35 MHz) per operator in the initial phase, moving towards the release of the entire 155 to 175 MHz within the core band before 2005. This is the spectrum designated for IMT-2000 at WARC-92 (World Administrative Radio Conference, 1992).

Of equal importance is the global release of the existing second-generation bands for second-generation GSM services. This would for example mean that spectrum between 862 and 960 MHz and the totality of the GSM1800 band, 1,710 to 1,880 MHz, should be released before 2002 for mobile applications as widely as possible.

3.2.5   GSM Frequencies

In principle, the GSM system can be implemented in any frequency band. However there are several bands where GSM terminals are, or will shortly be, available. Furthermore, GSM terminals may incorporate one or more of the GSM frequency bands listed below to facilitate roaming on a global basis.

- GSM400: 450.4 to 457.6 MHz paired with 460.4 to 467.6 MHz or 478.8 to 486 MHz paired with 488.8 to 496 MHz;
- GSM 850: 824 to 849 MHz (mobile transmit, base receive), 869 to 894 MHz (base transmit, mobile receive);
- GSM900: 880 to 915 MHz paired with 925 to 960 MHz;
- GSM1800: 1,710 to 1,785 MHz paired with 1,805 to 1,880 MHz;
- GSM1900: 1,850 to 1,910 MHz paired with 1,930 to 1,990 MHz.

In the above bands, MSs transmit in the lower frequency subband and BSs transmit in the higher frequency subband.

3.2.6   IMT-2000 (Third-Generation) Core Frequency Band

The third-generation frequencies for IMT-2000 were identified by the ITU in 1992 and appear as No. S5.388 of the Radio Regulations:

S5.388 The bands 1,885 to 2,025 MHz and 2,110 to 2,200 MHz are intended for use on a worldwide basis by administrations wishing to implement IMT-2000. Such use does not preclude the use of these bands by other services to which they are allocated. The bands should be made available for IMT-2000 in accordance with Resolution 212 (Rev.WRC-97).

Terrestrial IMT-2000 services will operate in the Frequency Division Duplex (FDD) mode in the bands 1,920 to 1,980 MHz paired with 2,110 to 2,170 MHz with MSs transmitting in the lower subband and BSs transmitting in the upper subband. The bands 1,885 to 1,920 MHz and 2,010 to 2,025 MHz are unpaired for Time Division Duplex (TDD) operation. Administrations and Regional Regulatory Telecommunication Organizations should be consulted concerning any specific national or regional arrangements for IMT-2000.

GSM association members will, subject to commercial, operational, and regulatory considerations, commence the introduction of third-generation IMT-2000 services from 2002.

3.2.7   IMT-2000 (Third-Generation) Extension Bands

At the May to June 2000 World Radiocommunication Conference (WRC-2000) the ITU identified additional IMT-2000 frequency bands.

In addition to the frequency bands currently designated for IMT-2000 in number S5.388 of the Radio Regulations (the IMT-2000 core-band) and those bands which in 2005 will be utilized by administrations for second-generation public land mobile services in their territories (e.g., GSM 900, 1,800, and 1,900), the frequency bands 698 to 806 MHz, 2,500 to 2,690 MHz, and 2,700 to 2,900 MHz should be nonexclusively designated for the use of IMT-2000 to provide for up to an additional 160 MHz of spectrum for IMT-2000 from 2005. Subject to commercial demand, the 160 MHz of additional spectrum should be available in its entirety prior to 2010. Existing second-generation bands (including GSM bands) 806 to 960 MHz, 1,429 to 1,501 MHz and 1,710 to 1,885 MHz should be confirmed as very long term IMT-2000 bands and studies commenced to pave their way for formal identification at a subsequent WRC. These were not be identified at WRC-2000 as IMT-2000 extension bands since they will continue to be extensively used for GSM (and other second-generation systems) for a long period of time for both GSM and in multimode GSM IMT-2000 terminals to supplement IMT-2000 coverage. All the bands (extension and existing second-generation bands) should be allocated in all ITU regions on a primary basis to the mobile service (if not already so allocated) in the ITU Radio Regulations.

Today's market estimates for mobile wireless personal communication and considering 3G and 4G proposals for wideband multimedia services show that spectrum resource management will remain an important aspect in the near future. As the number of transmitters in the system becomes large within some fixed chunk of available RF spectrum, the number of simultaneous connections (links) will become larger than the number of orthogonal signals the available bandwidth may provide.

3.3 RRM

By now, it should be clear to the readers that, in order to better utilize the scarce radio spectrum in mobile communication systems, the available radio resources (transmitter powers, channels, and BSs) must be used in the most efficient way. Efficient use of radio spectrum [3] is important from a cost-of-service point of view, where the number of BSs required to service a given geographical area is an important factor. A reduction in the number of BSs, and hence in the cost of service, can be achieved by more efficient reuse of the radio spectrum. All RRM schemes are designed using some model for the traffic. Most wireless systems today use circuit-switched speech as the main design model (e.g., GSM, CDMA one). Future wireless access systems are expected to carry both voice and data as well as a mixture of services with very different and often conflicting service requirements.

3.3.1 RRM Problem

We describe the RRM problem as formulated in [4]. Let us assume that M mobiles are served by BSs, numbered from the set $B = \{1, 2, 3,..., B\}$. Let us also assume that there are C orthogonal channel pairs numbered from the set $C = \{1, 2, 3,..., C\}$ available for establishing links between BSs and MTs. To establish radio links, the system has to assign to each mobile: (1) a BS from the set B, (2) a channel from the set C, and (3) a transmitter power for the access port and the terminal. This assignment is performed according to the resource allocation algorithm (RAA) of the wireless communication system. The assignment is restricted by the interference caused by the BSs and mobiles as soon as they are assigned a channel and when they start using it. Another common restriction is that BSs, in many cases, use only a subset of the available channels. Good allocation schemes will aim at assigning links with adequate signal-to-interference ratio (SIR) to as many (possibly all) mobiles as possible. Note that the RAA may well (should) opt for not assigning a channel to an active mobile if this assignment would cause excessive interference to other mobiles.

Let us now study the interference constraints on resource allocations in somewhat more detail. We now may compute the signal and interference power levels in all access ports and mobiles, given the link (power) gains, G_{ij}, between

BS i and MT j. For the sake of simplicity, we will here consider only wideband modulation schemes, which will make the link gains virtually independent of the frequency. Collecting all link gains in matrix form, we get a $(B \times M)$ rectangular matrix, the link gain matrix.

$$G = \begin{pmatrix} G_{11} & G_{12} & .. & .. & G_{1M} \\ G_{21} & G_{22} & .. & .. & G_{2M} \\ .. & & .. & .. & .. \\ .. & & .. & .. & .. \\ G_{B1} & G_{B2} & .. & .. & G_{BM} \end{pmatrix} \tag{3.1}$$

The link gain matrix describes the (instantaneous) propagation conditions in the system. Note that, in a mobile system, both the individual components (mobile motion) and the dimension of the matrix (traffic pattern) may vary over time. The task of the resource allocation scheme is to find assignments for which the SIR [signal-to-(interference + noise) ratio] is large enough [exceeds the threshold γo in as many links as possible (preferably all)]. This means that the following inequality must hold:

$$\Gamma_l \frac{P_j G_{ij}}{\Sigma P_m G_{im} \theta_{jm} + N} \geq \gamma o \tag{3.2}$$

where Γ_j denotes the SIR in the uplink (MT-to-BS) and downlink (BS-to-MT) of the connection and N denotes the receiver (thermal) noise power at the access port. P_j denotes the transmitter power used by terminal j. The quantity θ_{jm} is the normalized cross-correlation between the signals from mobiles j and m at the access port receiver. If the waveforms are chosen to be orthogonal (as in FDMA and TDMA) these correlations are either zero or one depending on whether the station has been assigned the same frequency (time slot) or not.

Note that it may be impossible to comply with all of the constraints [see (3.2)] for all of the mobiles M, in particular if M happens to be a large number. As the system designer, we may have to settle for finding resource allocation schemes that assign channels with adequate quality to as many mobiles as possible. The largest number of users that may be handled by the systems is a measure of the system capacity. Since the number of mobiles is a random quantity and the constraints [see (3.2)] depend on the link matrix, (i.e., on the relative position of the mobiles), such a capacity measure is not a well-defined quantity.

The classical approach of telegraphic theory is to use as capacity measure the maximal relative arrival rate of calls ρ for which the blocking probability (the probability that a newly arrived call is denied service) can be kept below some predetermined level. Due to the mobility of the MTs this is not an entirely satisfying measure. A call may be lost due to adverse propagation conditions. To include such phenomena into the above formulation, we would require detailed specification of call handling procedures (e.g., handling of new versus old calls, handoff procedures as a mobile moves from one access port to another). It may therefore be practical to choose a simpler and more fundamental capacity measure that will reflect the performance of the resource allocation scheme as such.

For this purpose, let us assume that, at some given instant, the RAA has succeeded in providing adequate links to a number of MTs Y out of the total number M. Y will of course be a stochastic variable. Let Z denote the remaining number of mobiles, for which the assignment fails, that is

$$Z = M - Y \qquad (3.3)$$

We define the assignment failure rate v as

$$v = \frac{E[Z]}{E[M]} = \frac{E[M]}{\omega A} \qquad (3.4)$$

In the last expression we have assumed the (active) MTs to be uniformly (2D Poisson) distributed over the service area A with ω mobiles per area unit. The quantity v is a measure of the extent to which the allocation scheme has been successful in providing the MTs with links of adequate quality. For moderate to large ωA, v is also a good approximation of the probability that a randomly chosen active mobile at some given instant is not provided with a channel. The instantaneous capacity $\omega^*(v_o)$ of a wireless system is the maximum allowed traffic load in order to keep the assignment failure rate below some threshold level v_0, that is,

$$\omega^*(v_o) = \{\max \omega : v \le v_o\} \qquad (3.5)$$

As we have seen above, it is a formidable problem to find the optimum resource allocation, that is, to determine for each MT the following:

1) A waveform assignment (determining the θ_{jm});
2) A BS assignment [of one or more (!) ports];
3) A transmitter power assignment that maximizes Y for a given link gain matrix.

No efficient general algorithm that is capable of doing such an optimal assignment for arbitrary link gain matrices and mobile sets is known [3]. Instead, partial solutions and a number of more or less complex heuristic schemes have been proposed (and are used in the wireless systems of today). These schemes are usually characterized by low complexity and by using simple heuristic design rules. The capacity ω^* achieved by these schemes is, as expected, often considerably lower than what could be expected by optimum channel assignment.

3.3.2 Channel Allocation and Assignment

An interesting subproblem of the general RRM problem that has attracted much attention in the literature is the choice and allocation of channels. Dividing the spectrum into a set of channels (i.e., identifying C) is the first step in RRM. A given radio spectrum (or bandwidth) can be divided into a set of disjoint or noninterfering radio channels. All such channels can be used to maintain an acceptable received radio signal. Many techniques, such as frequency division (FD), time division (TD), or code division (CD), can be used in order to divide a given radio spectrum into channels (i.e., a "channelization" of the spectrum). In FD, the spectrum is divided into disjoint frequency bands, while in TD the channel separation is achieved by dividing the usage of the channel into disjoint time periods called time slots. In CD, the channel separation is achieved by using different modulation codes. Also, dividing each frequency band of an FD scheme into time slots can use a combination of TD and FD. The major driving factor in determining the number of channels with certain quality that can be used for a given wireless spectrum is the level of received signal quality that can be achieved in each channel. Classical orthogonalization techniques, such as FDMA and TDMA are very common and used extensively in various systems. They are now being challenged, however, by nonorthogonal waveform systems of CDMA such as IS-95 [5–8].

Once the set C is known, the next problem is the assignment of waveforms to the different terminal access port links. This is known as the *channel assignment* (CA) problem. This assignment can be done in a lot of different ways depending on the amount and quality of the information available regarding the matrix G and the traffic situation (activity of different MTs). Another important issue is the time scale on which resource reallocation is feasible. In the radio and transmission subsystems, techniques such as deployment of time and space diversity systems, use of low-noise filters and efficient equalizers, and deployment of efficient modulation schemes can be used to suppress interference and to extract the desired signal. Cochannel interference caused by frequency reuse, however, is the most restraining factor on the overall system capacity in the wireless networks. The main idea behind channel assignment algorithms is to make use of radio propagation path-loss [1, 5] characteristics in order to minimize the CIR and hence increase the radio spectrum reuse efficiency.

Many of today's communication systems use a fixed assignment strategy to divide the resources among the users. This is known as static or fixed CA (FCA), and it mostly operates on a long-term basis. Based on average statistical information regarding G (i.e., large scale propagation predictions), frequencies are assigned to different BSs on a more or less permanent basis. Such a cell plan provides a sufficient reuse distance between BSs, providing a reasonably low probability of outage to low SIR [2]. The basic prohibiting factor in radio spectrum reuse is interference caused by the environment or other mobiles. Interference can be reduced by deploying efficient radio subsystems and by making use of channel assignment techniques to be discussed below. Significant overlap between cells is quite common in microcells and especially indoor picocells. This overlap makes it difficult to predict interference, and thus many resources are wasted when channels are assigned with FCA because the channels need to be divided into reuse clusters. Heterogeneities in the traffic load can also be taken care of by adapting the number of channels in each cell to the expected traffic carried by that BS. To minimize the planning effort, adaptive cell planning strategies (e.g., channel segregation [4]) have been devised using long-term average measurements of the interference and traffic to automatically allocate channels to the BS. These static (or "quasistatic") channel allocation schemes work quite well when employed in macrocellular systems with high traffic loads. In short-range (microcellular) systems and in multimedia traffic scenarios, static channel allocation schemes require considerable design margins to cope with the large variations in propagation conditions and traffic load. Large path-loss variations are countered with large reuse distances, unfortunately at a substantial capacity penalty. In the same way, microcellular traffic variations are handled by assigning excess capacity to handle traffic peaks. In recent years, however, to improve the trunking efficiency in such cases, the resources are shared more efficiently between the cells in a dynamic way [Dynamic CA (DCA)]. A number of dynamic channel selection strategies based on interference have been shown to outperform FCA [4]. We shall discuss the various CA schemes in detail next.

3.3.3 Schemes for CA

CA schemes can be divided into a number of different categories depending on the comparison basis. For example, when channel assignment algorithms are compared based on the manner in which cochannels are separated, they can be divided into three categories: (1) FCA, (2) DCA, and (3) hybrid channel assignment (HCA).

In FCA schemes, the area is partitioned into a number of cells, and a number of channels are assigned to each cell according to some reuse pattern, depending on the desired signal quality [6]. FCA schemes are very simple. As explained above, they do not adapt, however, to changing traffic conditions and user distribution. In order to overcome these deficiencies, DCA strategies have been introduced. In DCA, all channels are placed in a pool and are assigned to new calls as needed such that the threshold criterion of CIR (called CIR$_{min}$) is satisfied. At the cost of higher complexity, DCA schemes provide flexibility and traffic adaptability. Under high-load conditions, however, DCA strategies are less efficient than FCA. To overcome this drawback, HCA techniques were designed by combining FCA and DCA schemes.

In HCA, the total number of channels available for service is divided into fixed and dynamic sets. The fixed set contains a number of nominal channels that are assigned to cells as in the FCA schemes and, in all cases, are to be preferred for use in their respective cells. All users share the second (dynamic) set of channels in the system to increase flexibility. When a call requires service from a cell and all of its nominal channels are busy, a channel from the dynamic set is assigned to the call. CA schemes can be implemented in many different ways. For example, a channel can be assigned to a radio cell based on the coverage area of the radio cell and its adjacent cells such that the CIR$_{min}$ is maintained with high probability in all radio cells. Channels can also be assigned by taking the local CIR measurements of the mobile's and BS's receivers into account. That is, instead of allocating a channel blindly to a cell based on worst-case conditions (such as letting cochannels be located at the closest boundary), a channel can be allocated to a mobile based on its local CIR measurements [9, 10].

Whatever the adopted CA scheme may be, it can be implemented in either centralized or distributed fashion. In the centralized schemes, the channel is assigned by a central controller, whereas in distributed schemes a channel is selected either by the local BS of the cell from which the call is initiated or autonomously by the mobile. In a system with cell-based control, each BS keeps information about the current available channels in its vicinity. Here the channel availability information is updated by exchange of status information between BSs. Finally, in autonomously organized distributed schemes, the mobile chooses a channel based on its local CIR measurements without the involvement of a central call assignment entity. Obviously, this scheme has a much lower complexity at the cost of lower efficiency. It is important to note that channel assignment based on local assignment can be done for both FCA and DCA schemes.

3.3.3.1 FCA

In the FCA strategy, a set of nominal channels is permanently allocated to each cell for its exclusive use. Here, a definite relationship is assumed between each channel and each cell, in accordance with cochannel reuse constraints [6, 11–17]. The total number of available channels C in the system is divided into sets, and the minimum number of channel sets N required to serve the entire coverage area is related to the reuse distance σ as follows [12, 17]: $N = (1/3)\, \sigma^2$, for hexagonal cells. Here σ is defined as D/R, where R is the radius of the cell and D is the physical distance between the two cell centers [5]. N can assume only the integer values 3, 4, 7, 9,... as generally presented by the series, $(i + j)^2 - ij$, with i and j as integers [6, 11].

In the simple FCA strategy, the same number of nominal channels is allocated to each cell. This uniform channel distribution is efficient if the traffic distribution of the system is also uniform. In that case, the overall average blocking probability of the mobile system is the same as the call blocking probability in a cell. Because traffic in cellular systems can be nonuniform with temporal and spatial fluctuations, a uniform allocation of channels to cells may result in high blocking in some cells, while others might have a sizeable number of spare channels. This could result in poor channel utilization. It is therefore appropriate to tailor the number of channels in a cell to match the load in it by nonuniform channel allocation [18, 19] or static borrowing.¹ [20, 21]. In nonuniform channel allocation, the number of nominal channels allocated to each cell depends on the expected traffic profile in that cell. In contrast to static borrowing, channel borrowing strategies deal with the short-term allocation of borrowed channels to cells; once a call is completed, the borrowed channel is returned to its nominal cell. The channel borrowing schemes can be again divided into simple and hybrid schemes. Table 3.1 summarizes the channel borrowing schemes.

Table 3.1
Channel Borrowing Schemes

Category	Different Schemes
Simple channel borrowing	Simple borrowing
	Borrow from the richest
	Basic algorithm
	Basic algorithm with reassignment
	Borrow first available
Hybrid channel borrowing	Simple hybrid borrowing scheme
	Borrowing with channel ordering
	Borrowing with directional channel locking
	Sharing with bias
	Channel assignment with borrowing and reassignment
	Ordered dynamic channel assignment with rearrangement

From: [3].

3.3.3.2 DCA

Due to short-term temporal and spatial variations of traffic in cellular systems, FCA schemes are not able to attain high channel efficiency. To overcome this, DCA schemes have been studied during the past twenty years. In contrast to FCA, there is no fixed relationship between channels and cells in DCA. All channels are kept in a central pool and are assigned dynamically to radio cells as new calls arrive in the system [22, 23]. After a call is completed, its channel is returned to the central pool. In DCA, a channel is eligible for use in any cell provided that signal interference constraints are satisfied.

Because, in general, more than one channel might be available in the central pool to be assigned to a cell that requires a channel, some strategy must be applied to select the assigned channel [15]. The main idea of all DCA schemes is to evaluate the cost of using each candidate channel and to select the one with the minimum cost provided that certain interference constraints are satisfied. The selection of the cost function is what differentiates DCA schemes [15]. The selected cost function might depend on the future blocking probability in the vicinity of the cell, the usage frequency of the candidate channel, the reuse distance, channel occupancy distribution under current traffic conditions, radio channel measurements of individual mobile users, or the average blocking probability of the system [7]. Table 3.2 gives a list of the proposed DCA schemes.

3.3.4 Transmitter Power Control

Power control is another supplementary technique to handle RRM. Power control can serve various purposes:

- To suppress adjacent channel (cross-correlation) interference in nonorthogonal schemes;
- To minimize power consumption in order to extend terminal battery life;
- To control cochannel interference (in schemes with orthogonal waveforms).

As discussed in the previous section, the purpose of CA algorithms is to assign radio channels to mobile users such that a certain level of CIR is maintained at every MT. Hence, one can also use power control schemes to achieve the desired CIR level. This is why the selection of the proper transmitter power in MTs and BSs is another important issue that has attracted considerable attention in recent years.

In the resource allocation problem context, it can be shown that the maximum number of terminals is supported under a power control regime that

Table 3.2
Dynamic Channel Allocation Schemes

Different Schemes	Description
Centralized DCA	First available
	Locally optimized dynamic assignment
	Selection with maximum usage on the reuse ring
	Mean square
	Nearest neighbor
	Nearest neighbor + 1
	1-clique
Distributed DCA	Locally packing distributed DCA (LP-DDCA)
	LP-DDCA with ACI constraint
	Moving direction
CIR measurement DCA schemes	Sequential channel search
	Minimum Signal-to-Noise Interference Ratio
	Dynamic channel selection
	Channel segregation
One dimension systems	MINMAX
	Minimum interference
	Random minimum interference
	Random minimum interference with reassignment
	Sequential minimum interference

From: [3].

balances the CIR of all terminals that can be supported and shuts off the rest [6]. The purpose of different power control schemes is to find the trade-off between change of transmitted power level and interference. In a way, power control schemes try to reduce the overall CIR in the system by measuring the received power and increasing the transmitted power in order to maximize the minimum level of CIR in the system.

The power control, when combined with some DCA algorithm in a distributed way, can provide most of the capacity gain. Finding the optimum set of nonsupported terminals is a problem closely related to the design of DCA schemes. Distributed implementations and different implementational constraints [13, 14] have been studied. Results show that very robust near-optimum power control schemes can be devised at very low complexity. Performance results indicate that in static channel allocations substantial (greater than 100%)

capacity gains can be achieved using optimum power control. These gains are, of course, not additive with the gains obtained by DCA schemes. Preliminary results, however, regarding DCA combined with power control schemes show substantial capacity gains [15, 17].

3.4 Handoff Process

An issue that is closely related to CA techniques is the handoff mechanism, which is usually associated with the movement of MTs. Handoff is defined as the change of radio channel used by a MT. Handoff management enables the network to maintain a user's connection as the mobile terminal continues to move and change its access point to the network. This is performed in three steps: initiation, connection generation, and data flow control.

The cause of handoff may be poor radio quality due to a change in the environment or the movement of the MT. As MTs move about in the service area, the propagation and interference situation may change such that the same cell (or BS) on any waveform cannot support the terminal. In addition, new MTs may enter the service area requiring services, while others terminate their communication sessions. Since most of the basic resource allocation strategies deal mainly with static situations encountered on a short-term timescale, resource management schemes should be capable of handling these variations. In general, the handoff event is caused by the radio link degradation or is initiated by the system that rearranges radio channels in order to avoid congestion. Handoff is extremely important in any mobile network because of the cellular architecture employed to maximize spectrum utilization. Handoff can be classified into network-controlled (or hard) handoff, and mobile-controlled (or soft) handoff. We shall discuss them now.

3.4.1 Network-Controlled Handoff (Hard Handoff)

In network-controlled handoff, when a mobile user moves to the edge of the cell boundary, the network invokes the handoff process. Since it is either all or nothing, the transition is a "hard" one. The user experiences an interruption caused by frequency shifting. Handoff initiation occurs dependent on different criteria and strategies used. The most common criteria are (1) relative signal strength, (2) relative signal strength with threshold, (3) relative signal strength with hysteresis, and (4) relative signal strength with hysteresis and threshold (see Figure 3.3).

Hard handoff may be intracell or intercell, depending upon whether the new radio channel is with the same BS (intracell handoff) or with a new BS (intercell handoff). In the first case, it is a waveform reallocation within the same

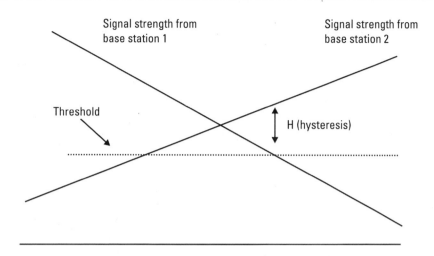

Figure 3.3   Handoff scenario between cells 1 and 2 based on hysteresis.

cell. In the second case, it is a relay from a cell to one of its neighboring cells, which in principle involves re-execution of the basic channel allocation scheme. The intercell handoff mechanism has often been modeled as a macrodiversity scheme where the terminal is assigned to the access cells with the highest received signal level. In high-density wireless systems, the coverage areas of the access cells overlap to a large extent. Low signal levels are rarely a problem since, normally, several access cells provide sufficient signal levels. In these cases, variations in interference, not cell boundary crossings, are the most probable cause of a handoff.

Understandably, intercell handoff is more crucial than intracell handoff. Intercell handoff transfers an ongoing call from one cell to another cell as a users move toward the coverage area of the neighboring cell. For example, the mobile subscriber might cross cell boundaries and move to an adjacent cell while the call is in process. In this case, the call must be handed off to the neighboring cell in order to provide uninterrupted service to the mobile subscriber. If adjacent cells do not have enough channels to support the handoff, the call is then blocked. Each handover requires network resource to route the call to the next BS. If handover does not occur quickly, the quality of service (QoS) may degrade below an acceptable level, and the connection will be dropped.

3.4.2   Mobile-Controlled Handoff (Soft Handoff)

Soft handoff is the ability of an MT to select between the instantaneous received signals from different BSs. Soft handoff allows an MT to communicate with multiple BSs simultaneously. It is one of the advantages of CDMA over TDMA.

Soft handoff is an effective way to increase the capacity, reliability, and coverage range of CDMA systems.

In TDMA or Advanced Mobile Phone System (AMPS), due to spectrum reuse, a given slot on a given frequency channel cannot be used by neighboring cells. So when an MT that is in a call moves from one cell to another, at a certain point it has to switch between cells. In AMPS and TDMA, it will be commanded by the system to change frequencies all at once (hard handoff, as discussed above). On the contrary, in CDMA, all the cells operate on the same frequency. The MT still has a single RF receiver, which converts RF down to baseband, but behind that it has a rake receiver with multiple fingers. Since all the cells operate on the same frequency, the single RF receiver picks up all of those that are within range. The MT then assigns fingers from the rake receiver to various signals, and these are added together to create the full signal that the MT utilizes. Sometimes these are multiple paths from the same cell. For instance, if there is a direct route from the cell to the MT, and in addition the signal travels to a large building and reflects off it before reaching the MT, then the MT can utilize both of these signals for additional clarity. We have already discussed this as a multipath phenomenon, in Section 3.1, which essentially degrades TDMA and AMPS performance. But soft handoff is even more useful when the MT is about halfway between two cells. While in a call, the MT is not only handling the transport of data back and forth to the cell, but it is also actively looking for other cells. When the MT finds a cell with good signal strength (on the same frequency, remember) it will inform the cell system. The cell system might decide, at that point, to route the call through both cells simultaneously. The specification actually permits an MT to talk to six cells at once, though no MT currently in existence has this capability.

There are two key requirements for realizing soft handoff:

1. *Data distribution and selection:* Separate copies of the same data need to be sent via multiple BSs to the same MT (or in the reverse direction, from an MT to multiple BSs).

2. *Data content synchronization:* Pieces of data arriving from multiple BSs to an MT at the same time should be copies of the same data in order for the MT's radio system to correctly combine these copies into a single copy. In the reverse direction, only one copy of the data sent by the MT to multiple BSs should be selected for delivery to the destination.

So when a CDMA phone in a call moves from one cell to another, the handoff process happens in multiple steps. First the phone notices the second cell, and the cell system begins to carry the call on both cells. As the phone continues to move, eventually the signal strength from the cell the phone is moving away from will drop to the point where it is no longer useful. Again, the phone will inform the cell system of this fact, and the system will drop the original cell. Thus it is not an all-or-nothing transition, which is why it is called a soft handoff. Soft handoff has proven to offer superior performance with respect to hard handoff in terms of maximizing the uplink capacity of wireless cellular networks. This capacity increase comes from the macroscopic diversity exploited by soft handoff, whereby signals received by several BSs are combined, as opposed to the selection of the best signal performed in hard handoff. The performance of soft handoff, however, is very sensitive to the settings of the parameters involved in its implementation. Therefore, an optimization study of soft handoff with respect to the optimal parameter settings is essential for its successful implementation in wireless networks.

3.4.3 Handoff Prioritizing Schemes

In systems where the cell size is relatively small (microcellular systems), the handoff procedure has an important effect on the performance of the system. An important issue, here, is to limit the probability of forced call termination, because, from the point of view of a MU, forced termination of an ongoing call is less desirable than blocking a new call. Therefore, the system must reduce the chances of unsuccessful handoffs by reserving some channels explicitly for handoff calls. For example, handoff prioritizing schemes are channel assignment strategies that allocate channels to handoff requests more readily than new calls. Handoff prioritizing schemes provide improved performance at the expense of a reduction in the total admitted traffic and an increase in the blocking probability of new calls. The simplest way of giving priority to handoff calls is to reserve some channels for handoff calls explicitly in each cell. This scheme is referred to as the cutoff priority scheme (CPS) [19, 24, 25] or the guard channel scheme [26, 27]. Other prioritizing schemes allow either the handoff to be queued [24, 26] or new calls to be queued [27] until new channels are obtained in the cell. Several variations of the basic cutoff priority scheme are available with the queuing of handoff requests or of new call requests [25–27]. The guard channel concept can be used in FCA or DCA schemes. Here, guard channels are not assigned to cells permanently; instead, the system can keep a collection of channels to be used only for handoff requests, or have a number of flexible channels with associated probabilities of being allocated for handoff requests.

3.5 Managing Resource Allocation

As discussed in the previous sections, the challenges in the wireless networks are to guarantee QoS requirements while taking into account the RF spectrum limitations and radio propagation impairments. As the demand for wireless service increases, managing radio resource becomes more complicated. Figure 3.4 depicts the generic functional blocks used for RRM in a typical cellular system. This shows that CA strategy has to be supported by handoff strategy, power control, and call admission control (CAC) in order to entail a balanced RRM. We have already discussed CA, handoff, and power control above. Here we will dwell on CAC and QoS issues to complement our earlier discussions.

3.5.1 CAC

CAC denotes the process of making a decision for every new call admission according to the amount of available resource versus users' QoS requirements, and the effect upon the QoS of the existing calls imposed by each new call. Whenever a new MT (either a new request for service or an intercell handoff) arrives in a BS, the RRM system has to decide if this particular MT may be allowed into the system. An algorithm making these decisions is called a CAC algorithm. Since the precise MT population and gain matrix may not be tracked exactly at all times, and due to the complexity of RRM algorithms, determining the success of an admission decision may not be possible beforehand without physically executing the admission itself. CAC can be divided into two groups,

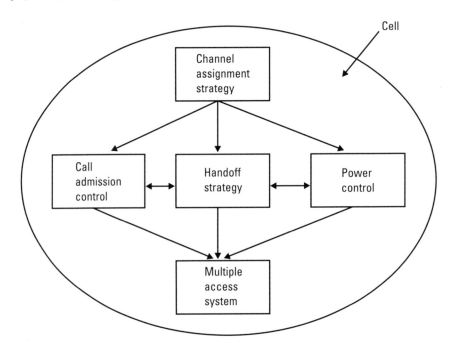

Figure 3.4 Generic function blocks of CAC.

interference-based CAC and user-based CAC. In interference-based algorithms, admission control is based on the CIR parameter, while user-based algorithms are dependent on the number of channels available.

Traditional approaches involving static channel allocation normally use simple thresholding strategies on the available channels in each cell. This traditional CAC algorithm is known as the guard channel algorithm. The guard channel is a concept based on reserving a fraction of channels for the purpose of handoff by considering only the status for a local cell. Here, BSs are assigned a fixed set of channels that provide a certain low-outage probability. Such a system is called blocking-limited. Whenever a call arrives, the system checks whether or not there are channels available. New arriving calls are admitted up to the point at which there is only some small fraction of the resource remaining. This spare capacity is reserved for calls entering a cell due to an intercell handoff.

In systems using dynamic channel allocation or random allocation, there is no clear limit on the number of channels or waveforms that can be used. In such interference-limited systems, the feasibility of admitting new users will depend on the current interference situation. This is particularly complicated in systems utilizing CIR-balancing power control by the fact that terminals already active will react to the admission of a new terminal by adjusting (raising) their transmitter powers. When several access cells provide sufficient CIR, the system is also able to handle traffic variations by means of load sharing. However, providing guaranteed service (in terms of continuity in channels) to the MTs in a large (possibly global) wireless system, coupled with mobility management, is a formidable task indeed.

There are also some nontraditional methods that have been proposed recently. They are based on adaptive techniques, where channels are allocated and reserved in a dynamic way using teletraffic analysis, prediction of injected traffic, and prediction of MT movement. In some prediction schemes, it is sufficient to take care of the radio resource that the mobile will need in the predicted location. In general, the resource reservation mechanism consists of two parts: some of the bandwidth reserved in the next cell the mobile is likely to visit, and a common pool of dynamically adjusted bandwidth used to accommodate other unpredicted flows. The next cell is predicted based on the mobility pattern observed in various cells. An approach that utilizes reservation of bandwidth along the predicted path of the mobile, or along some neighbors ofNthat path, is based on the concept of shadow clustering. The fundamental idea of the shadow-cluster concept is that, as an active MT travels to another cell, the region of influence also moves, following the active MT to its new location. The BSs currently being influenced are said to form a shadow cluster, because the region of influence follows the movement of the active MT like a shadow.

Other similar approaches include schemes that rely on the extended location information (from adjacent cells) to make the CAC decision. The bandwidth reservation could be estimated based upon the history of the nominal cell and neighboring cells. The queuing of new calls and handover requests is another approach that reduces the blocking probability and probability of forced termination. There is a trade-off between the increase in service quality and the corresponding decrease in total carried traffic. The queuing of a handoff request is more sensitive to delay in service than the queuing of a new call, leading to queuing of new calls rather than handoff calls. One of the key points of using queuing in CAC is that service differentiation can be managed with modified queuing disciplines. Instead of a first in–first out (FIFO) queuing strategy, other prioritized queuing disciplines can be used to maintain priority levels in each service class.

3.5.2   QoS

As we move towards the next-generation mobile systems, the need for improving coverage, system capacity, and service quality becomes more and more important. Different service classes have different QoS requirements. Future services will also have widely varying QoS requirements. The network must be able to handle these requirements in order to satisfy end users without wasting network resources. Moreover, to have any meaning, QoS provisioning must be end-to-end (i.e. from service to terminal). The radio interface is the scarcest resource in the mobile network and is a rather hostile environment, being error-prone and subject to radio propagation conditions that can vary over time. Consequently, effective QoS management in the radio network layer of the mobile network is a must. RRM will be the major differentiator between the overall QoS provisioning offered by different operators' networks.

To satisfy these requirements, the network must have enough capacity to serve arriving calls and flexible mechanisms to guarantee certain services. Connections are allocated to the system and cell layer that provide the requested QoS for the requested service most effectively. In this way, network resources are not wasted providing excess QoS to those services that do not require it, while higher quality resources are reserved for those services that require them. Within each system, the radio parameters will be set according to the QoS requirements of the radio service. These include the selection of channel, coding rate, and retransmission scheme for example. Conventionally, in a single layer system, reserving a number of channels in advance, known as guard channels, often does support prioritized handoff. Allocation of channels can either be dynamic or static, depending on the traffic pattern. Queuing is another method favoring handoff calls. By queuing new call arrivals, available bandwidth can be dedicated to handoff calls.

With the traditional single layer architecture, it is difficult to implement a flexible mechanism to guarantee QoS. Moving from a single cell layer towards a hierarchical multilayer system (as shown in Figure 3.5) can solve many problems related to the single cell architecture. Different cell layers (logically) can be dedicated to different user groups based on their mobility and traffic demands. For example, fast movers may be assigned to the macrocell layer while slow movers may be assigned to microcell layers. The use of microcells in hierarchical layered structures improves transmission quality, provides flexibility in traffic coverage, and significantly increases the system capacity. Overlaying macrocells enhances radio coverage at a low cost. In hierarchical cellular systems (HCS), providing QoS can be improved by having freedom in regulation of injected traffic λ_{ij} (see Figure 3.5) between the cell layers. If, say, one slow mover cannot be handed off to a microcell, he or she may be temporarily accommodated in the macrocell. On the other, a fast mover may be given a microcell for a while before being accommodated in the macrocell. Furthermore, we have seen that CAC is one method to manage radio resource in order to adapt to traffic variations. Guard channels and queuing can also be adopted in HCS, giving rise to a flexible CAC mechanism. The challenges, however, are how to (1) design a simple (fast) and efficient CAC scheme that can support multimedia applications and high-speed users, (2) design queuing strategies for handoff calls in a macrocell (queuing in a microcell is not preferred because of the cell size), and (3) investigate the possibility of a different medium access strategy in different hierarchical layers. Different layers logically represent different traffic patterns.

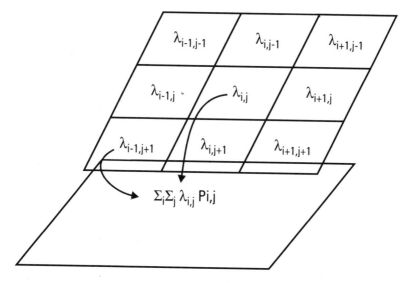

Figure 3.5 HCS.

3.6   Emerging RRM Techniques

Spectral reuse, system complexity, and coverage are some of the most important issues in a cellular architecture. Cellular systems with higher spectral efficiency give more complex RRM systems as the numbers of channels assigned to a cell becomes insufficient to support the required number of users. At this point, we need to improve system capacity using efficient RRM techniques. Techniques, such as cell splitting, sectoring, and adaptive antennas that could be used to expand the system capacity, are discussed in this section.

3.6.1   Cell Partitioning

A common method of increasing the spectral efficiency in cellular systems is reuse partitioning (RUP). In RUP, each cell in the system is divided into two or more concentric subcells or zones as shown in Figure 3.6. The inner zone, closest to the BS, requires lower power levels to achieve the desired CIR than the outer zone, so the minimum reuse distance for the inner zone become smaller than for the outer zone which leads to higher spectrum efficiency. RUP can be divided into fixed and adaptive techniques. Another variant of partitioning, called multiple channel bandwidth system (MCBS), can be utilized to increase the spectral efficiency. In MCBS, the cell is also divided into two or more concentric subcells. To achieve the desired CIR, the inner zone requires less

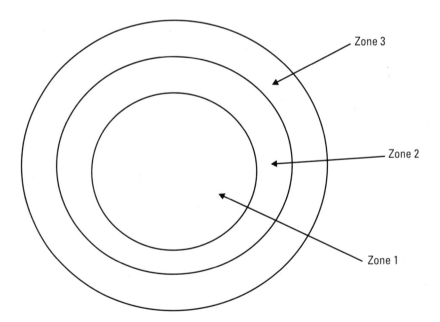

Figure 3.6   Concentric subcells.

bandwidth per user than the outer zone, thereby making more channels available in the inner zone. Thus, instead of utilizing the same amount of bandwidth per user throughout the whole cell, the MCBS can be used to increase the number of channels in a cell. A bandwidth is divided into a number of channels (i.e., bandwidth per user).

3.6.2 Multilayered Architecture

Underlay-overlay (overlay-underlay) architecture can also be used to increase system capacity, performance, and coverage. Normal cellular systems follow overlay-underlay structure where clusters of cells are usually grouped into a location area. Underlay-overlay differs from overlay-underlay in that a tighter reuse factor is used within the former. Figure 3.7 illustrates the hierarchical architecture of underlay and overlay systems. For example, a single overlay macrocell can be divided into two underlay clusters of microcells in an underlay-overlay system, hence increasing the number of channels in the whole system.

3.6.3 Software Radio

Joseph Mitola III (http://ourworld.compuserve.com/homepages/jmitola), a consulting scientist, coined the term software radio in 1991 to signal the shift from hardware intensive digital radios of the 1980s to the multiband,

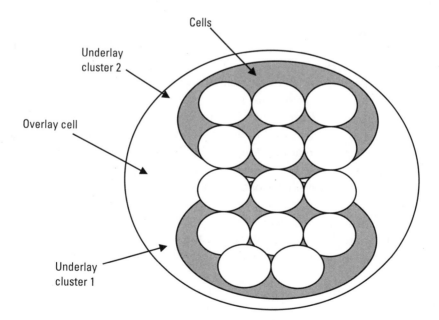

Figure 3.7 Hierachical architecture.

multimode, software-defined radios of 2000 and beyond. According to him, in the future, software radio will give an MT many different radio access interfaces without increasing the terminal complexity or cost.

As described in Chapter 1, radio network architectures have already evolved from early point-to-point and relatively chaotic peer networks (e.g., push-to-talk mobile military radio networks) toward more hierarchical structures with improved service quality (Cellular and PCS networks). In addition, channel data rates continue to increase through multiplexing and spectrum spreading. As communications technology continues its rapid transition from analog to digital, more functions of contemporary radio systems are implemented in software—leading toward software radio [28]. With software radio technology, MTs can access different networks according to their services. In a multiple-hierarchy application, a single radio unit, typically an MT, participates in more than one network hierarchy. A software radio terminal could, for example, operate in a GSM-based PCS network, an AMPS network, and a future satellite mobile network.

Technically speaking, a software radio is a radio whose channel modulation waveforms are defined in software. That is, waveforms are generated as sampled digital signals, converted from digital to analog via a wideband digital to analog converter (DAC) and then possibly upconverted from intermediate frequency (IF) to RF. The receiver, similarly, employs a wideband analog to digital converter that captures all of the channels of the software radio node. The receiver then extracts, downconverts and demodulates the channel waveform using software on a general-purpose processor. Software radios employ a combination of techniques that include multiband antennas and RF conversion; wideband ADC and digital to analog conversion, and the implementation of IF, baseband, and bitstream processing functions in general-purpose programmable processors. The resulting software radio, or software-defined radio (SDR), in part extends the evolution of programmable hardware, increasing flexibility via increased programmability. And in part it represents an ideal that may never be fully implemented but that nevertheless simplifies and illuminates trade-offs in radio architectures that seek to balance standards compatibility, technology insertion, and the compelling economics of today's highly competitive marketplaces. It is pertinent to mention here that there is a worldwide SDR Forum (http://www.sdrforum.org), which is an open, nonprofit corporation dedicated to supporting the development, deployment, and use of open architectures for advanced wireless systems. The Forum membership is international and is growing enormously.

The use of a software radio that can program a specific waveform for use in different wireless settings will make it possible to adapt the link to changing conditions. Such an approach is also expected to facilitate the management of QoS. In a region with many different independent radio access networks, one cannot explore the optimal use of available frequency resource and distribution of users within different radio access networks as one can in a centralized system. The idea is to integrate different distributed radio access networks to a centralized network architecture. With a centralized system, we can gain increasing trunking capacity and QoS in the region. The terminal reconfigurability provided by SDR technology introduces flexibility in spectrum management when different heterogeneous radio systems exist in the same geographical area. Considering development of and experimentation with wireless multimedia to be of great importance, initiatives in this area would be extremely useful. Development of such a software radio would help develop commercial products that can handle several services (e.g., AMPS, DAMPS, GSM, PCS, CDMA) in a single region. Trying out new schemes and ideas would also consume very little time with software radio. Software radio is now being enhanced to produce *cognitive radio,* an emerging topic within software radio. It refers to that class of software radio that employs model-based reasoning and at least a chess-program level of sophistication in using, planning, and creating radio etiquettes.

3.7 Integrated RRM

Future mobile networks will not consist simply of one radio access technology, such as WCDMA or EDGE, but will contain several different technologies as shown in Figure 3.1. The structure of operators' networks is becoming more complex, including multiple radio technologies in hierarchical cell structures; macrocell, microcell, and picocell layers. It is even possible that the same radio technology will be deployed on several different frequency bands. Simply speaking, the future is a hybrid (multiradio) environment. This will allow the networks to benefit from the individual coverage and capacity characteristics of each one, resulting in the most economic solution with the highest levels of service. In such a network, a multimode terminal will have the possibility of using a wide range of different resources. The motivation for hybrid networks arises from the fact that no one technology or service can provide ubiquitous coverage, and it will be necessary for an MT to employ attachment to different networks to maintain connection during the service time.

The emergence of a variety of mobile data services with variable coverage, bandwidth, and handoff strategies has attracted tremendous attention to the need for roaming among different radio access networks in a hybrid network architecture. The capabilities of network operators6 current infrastructure can be evolved and form part of this multiradio hybrid network, maximizing the usage of existing network investments. For this vision to work, these different technologies must be seamlessly integrated to form a single access network so that the end user will be unaware of the access technology being used. Similarly, from the operation and management point of view, these different technologies must be fully integrated to form a single network. Managing these technologies separately will be expensive, resulting in low resource usage and poor network quality. Moreover, hybrid networks will take advantage of the properties in the existing networks to provide a wider coverage and to serve all types of service classes. In a network with a mixture of resources (different systems and different layers) and offering a mixture of different services, however, it is vital to provide the optimum radio bearer for each service, based on the QoS requirements of the service. Overprovisioning QoS to those services or users that do not require it will waste network resources and should be avoided.

One solution to exploit hybrid networks is to provide a single integrated handoff and CAC strategy. For example, Figure 3.8 depicts the model of an integrated access system with a virtual RRM access node. This virtual node acts like a control node in a centralized system. The challenge, however, lies in the design of a virtual node considering various factors, such as the traffic generated in the system, the cell crossing rate (mobility analysis), and traffic overflowing amongst different radio access networks.

It can provide extra capacity to the network, resulting in higher end-user average bit rates and lower blocking. In particular, it can provide (1) load sharing, congestion control, and interference distribution; (2) simplified interworking in a multivendor and multisystem environment; (3) unified radio bearer

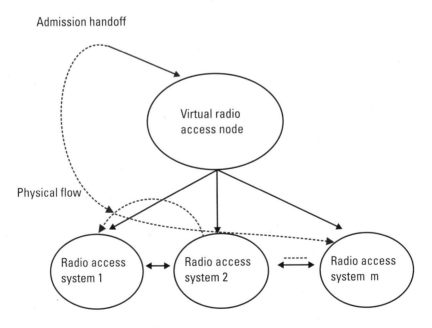

Figure 3.8 Model of a virtual access node.

QoS management; and (4) easier operability. Another important issue is to ensure that the terminal is directed to the optimum bearer, ensuring both that network resources are efficiently used and that the end user receives the required level of service (see Figure 3.9). Clearly, to get the best out of this mix of resources, some form of overall resource management is needed. This will manage all the network resources efficiently as a single pool and will offer the maximum utilization of all the operators' air interfaces, leading to higher user bit rate and lower blocking. Such a scheme can reduce unnecessary air interface signaling, core network signaling, and intersystem measurements. Basic RNC- and BSC-based intersystem handover algorithms can begin to exploit some of this gain, but cannot provide the optimum solution. Integrated RRM is the right approach towards the optimum solution to exploit this gain.

3.8 Summary

The rapid growth in demand for mobile communication has led to intense research and development efforts towards a new generation of cellular systems. The new system must be able to provide QoS, support a wide range of services while improving the system capacity. Efficient utilization of the scarce spectrum allocation for cellular communications is certainly one of the major challenges in cellular system design.

In radio and transmission subsystems, techniques such as deployment of time and space diversity systems, use of low noise filters and efficient equalizers, and efficient modulation schemes can be used to suppress interference and extract the desired signal. Cochannel interference caused by frequency reuse, however, is the most restraining factor on the overall system capacity in wireless

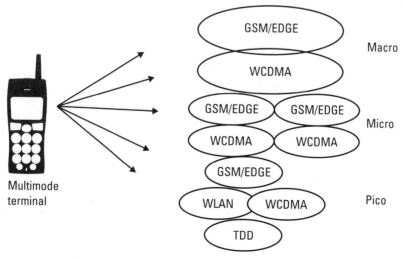

Figure 3.9 Integrated RRM model.

networks, and the main idea behind channel assignment strategies is to make use of radio propagation path-loss characteristics in order to minimize the CIR and hence increase the radio spectrum reuse efficiency.

In conclusion, a few remarks follow regarding what we believe to be future trends in radio resource allocation. Wideband and mixed rate traffic in small cell environments will exhibit a very large peak to average capacity demands. DCA (statistical multiplexing) will provide even larger capacity gains in these situations. Conventional single cell traffic multiplexing and averaging will surely not be sufficient, making dynamic spatial resource reuse of paramount importance [8]. Small cell systems allow greater spectral reuse and larger capacity, but induce an increasing number of handoffs, while overlaying cells provide coverage and service for high-speed users. Hierarchical cell architectures can therefore serve all types of user classes with different services and speed. The system performance characteristics include blocking probability, handoff blocking probability, forced termination of ongoing calls, and total carried traffic.

Traditionally we consider the frequency spectrum to be the resource to be shared. Since theoretically, there does not exist any upper limit on the capacity that can be provided (with a dense enough infrastructure), it is important that we widen the resource management perspective. Parameters such as infrastructure density costs and terminal power consumption play important roles. One could easily identify trade-offs such as where the signal-processing load should be put in a wireless system—in the terminal where power is scarce or in the fixed infrastructure. The key question here is: Should the access port (cell) infrastructure be very dense (and costly) allowing for "dumb," cheap, low-power terminals or should terminals be more complex allowing for the rapid deployment of a cheap infrastructure at the expense of battery life and terminal cost?

Future systems are expected to require much higher data rates than current systems. In third-generation wide-area personal communication systems (such as UMTS and FPLMTS) data rates in the range of 64 Kbps to 2 Mbps are coming [8]. Since the required transmitter power increases linearly with the bandwidth, high-speed radio access will have only a very limited range. This clearly has repercussions on the economics of such systems: either the investment will have to be heavy in a dense ubiquitous infrastructure, or be limited to covering only certain areas where users will require extensive bandwidth requirements. The design and performance of RRM algorithms is not affected much by the increase in bandwidth per se. In fact, much of the on-air signaling required by many of the adaptive schemes will, relatively speaking, occupy a smaller fraction of the available bandwidth. Increasing the infrastructure density with more BSs will clearly cause an increase in complexity in the RRM algorithms.

CDMA has been widely accepted as the major multiple access scheme in third-generation mobile communication systems [8]. WCDMA and its hybrid associate time-division CDMA are key elements of the IMT-2000 framework of standards. Since the beginning of the 1990s, there has been enormous research activity in analysis of the soft (i.e., interference-limited) capacity of these CDMA-based systems. Optimal usage of the soft capacity to provide, maintain, and guarantee QoS for different service classes is now becoming a very important issue. Therefore, interest in radio resource allocation has recently increased. RRM schemes (primarily for CDMA-based systems) should be flexible, support traffic services with various QoS requirements, minimize call or session blocking and dropping probabilities, and have acceptable radio resource utilization.

References

[1] Neskovic A., N. Neskovic, and G. Paunovic, "Modern Approaches in Modeling of Mobile Radio Systems Propagation Environment," *IEEE Communications Survey,* Third Quarter, 2000.

[2] Lopez, A. R., "Application of Wedge Diffraction Theory to Estimating Power Density at Airport Humped Runway," *IEEE Trans. on Antennas and Propagation,* June 1987, pp. 689–714.

[3] Katzela, I., and M. Naghshineh, "Channel Assignment Schemes for Cellular Mobile Telecommunication Systems: A Comprehensive Survey," *IEEE Personal Communication,* June 1996, pp. 10–31.

[4] Zander, J., "Radio Resource Management in Future Wireless Networks: Requirements and Limitations," *IEEE Communications Magazine,* Aug. 1997, pp. 30–96.

[5] Lee, W. C. Y., "New Cellular Schemes for Spectral Efficiency," *IEEE Trans. on Vehicular Technology,* 1987, pp. 188–192.

[6] Lee, W. C. Y., Mobile Cellular Communication Systems, 1989.

[7] Tekinay, S., and B. Jabbari, "Handover and Channel Assignment in Mobile Cellular Networks," *IEEE Communications Magazine,* 1991.

[8] Jorguseski, L., J. Farserotu, and R. Prasad, "Radio Resource Allocation in Third-Generation Mobile Communication Systems," *IEEE Communications Magazine,* Feb. 2001, pp. 117–123.

[9] Zander, J. "Asymptotic Bounds on the Performance of a Class of Dynamic Channel Assignment Algorithms," *IEEE JSAC,* 1993, pp. 926–933.

[10] Chuang, J .C. I., "Performance Issues and Algorithms for Dynamic Channel Assignment," *IEEE JSAC,* 1993.

[11] Kahwa, T. J., and N. Georganas, "A Hybrid Channel Assignment Scheme in Large Scale Cellular-Structured Mobile Communication Systems," *IEEE Trans. on Communication,* 1978, pp. 432–438.

[12]   Sin, J., and N. Georganas, "A Simulation Study of a Hybrid Channel Assignment Scheme for Cellular Land-Mobile Radio Systems with Erlang-C Service," *IEEE Trans. on Communication*, 1981, pp. 143–147.

[13]   Cox, D., and D. O. Reudink, "Increasing Channel Occupancy in Large-Scale Mobile Radio Systems: Dynamic Channel Reassignment," *IEEE Trans. on Vehicular Technology*, 1973, pp. 218–222.

[14]   Cox, D., and D. O. Reudink, "A Comparison of Some Non-Uniform Spatial Demand Profiles on Mobile Radio System Performance," *IEEE Trans. on Communication*, 1972, pp. 190–195.

[15]   Cox, D., and D. O. Reudink, "Dynamic Channel Assignment in Two Dimension Large-Scale Mobile Radio Systems," *Bell Sys. Tech. J.*, Vol. 51, 1972, pp. 1611–1628.

[16]   Schiff, L. "Traffic Capacity of Three Types of Common User Radio Communication Systems," *IEEE Trans. on Communication Tech.*, 1970, pp. 12–21.

[17]   Jakes, W. C., Jr., *Microwave Mobile Communications*, New York: Wiley, 1974.

[18]   Zhang, M., and T. S. Yum, "The Non-Uniform Compact Pattern Allocation Algorithm for Cellular Mobile Systems," *IEEE Trans. on Vehicular Tech.*, 1991, pp. 387–391.

[19]   Oh, S. H., et al., "Prioritized Channel Assignment in a Cellular Radio Network," *IEEE Trans. on Communication*, 1992, pp. 1259–69.

[20]   Anderson, L., "A Simulation Study of Dynamic Channel Assignment Algorithms in High Capacity Mobile Telecommunications System," *IEEE Trans. on Vehicular Tech.*, 1973.

[21]   Engel, J. S., and M. Peritsky, "Statistically Optimum Dynamic Server Assignment in Systems with Interfering Servers," *IEEE Trans. on Vehicular Tech.*, Vol. VT-22, 1973, pp. 203–209.

[22]   Okada, K., and F. Kubota, "On Dynamic Channel Assignment Strategies in Cellular Mobile Radio Systems," *IEICE Trans. Fundamentals*, 1992, pp. 1634–1641.

[23]   Okada, K., and F. Kubota, "On Dynamic Channel Assignment in Cellular Mobile Radio Systems," *Proc. IEEE Int. Symp. on Circuits and Systems*, 1991, pp. 938–941.

[24]   Hong, D., and S. Rappaport, "Traffic Modeling and Performance Analysis for Cellular Mobile Radio Telephone Systems with Prioritized and Nonprioritized Handoff Procedures," *IEEE Trans. on Vehicular Tech.*, 1986, pp. 77–92.

[25]   Tekinay, S., "A Measurement-Based Prioritization Scheme for Handovers in Mobile Cellular Networks," *IEEE JSAC*, Vol. 1, 1992, pp. 1343–1350.

[26]   Posner, C., and R. Guerin, "Traffic Policies in Cellular Radio that Minimize Blocking of Handoffs," *ITC-II*, 1985, pp. 2.4B.2.1–2.4B.2.5.

[27]   Guerin, R., "Queuing Blocking System with Two Arrival Streams and Guard Channels," *IEEE Trans. on Communication*, 1988, pp. 153–163.

[28]   Burns, P., *Software Radio Systems Design*, Norwood, MA: Artech House, 2002.

4

Location Management

Portable computers and communication devices with wireless connection to the network are changing the way people think about and use computing and communication. These wireless devices can communicate with each other even though the user is mobile. People carrying a mobile computer will, therefore, be able to access information regardless of time and current position. For example, they will be able to receive and send e-mail from any location or receive current information about local traffic, bus, and train services. But, location management will be an important problem in these situations because wireless devices can change location while connected to the network. New strategies must be introduced to deal with the dynamic changes of a mobile device's network address. A detailed description of the means and techniques for user location management in present cellular networks is addressed in this chapter.

The ability to change locations while connected to the network creates a dynamic environment. This means that data, which is static for stationary computing, becomes dynamic for mobile computing. A stationary computer, for example, is permanently attached to the nearest server, while mobile computers need a mechanism to determine which server to use. There are a few questions that must be answered when looking at a location management scheme. What happens when a mobile user changes location? Who should know about the change? How can you contact a mobile host? Should you search the whole network or does anyone know about the mobile users moves? Location management schemes are essentially based on users' mobility and incoming call rate characteristics. Two basic mechanisms to determine a mobile terminal's current location are: location update (or registration) and paging. The network mobility process has to find a balance between these two basic procedures. The location update procedure allows the system to keep the user's location knowledge, more or less accurately, in order to be able to find him or her, in case of an incoming call. Location registration is also used to bring the user's service profile near its location and allows the network to provide the user with his or her services rapidly. The paging process achieved by the system consists of sending paging messages in all cells where the MT could be located. Location management methods are broadly classified into two groups. The first group includes all methods based on algorithms and network architecture, mainly on the processing capabilities of the system. The second group contains the methods based on learning processes, which require the collection of statistics on subscribers' mobility behavior, for instance.

For location management purposes, cells in a cellular network are usually grouped together into location areas (LAs) and paging areas (PAs). An LA is a set of cells, normally (but not necessarily) contiguous, over which an MS may roam without needing any further location updates. In effect, an LA is the smallest geographical scale at which the location of the MS is known. A PA is the set of cells over which a paging message is sent to inform a user of an incoming call. A network must retain information about the locations of endpoints in the network in order to route traffic to the correct destinations. In cellular networks, endpoint mobility within a cell is transparent to the network, and hence location tracking is only required when an endpoint moves from one cell to another. In location management, it is important to differentiate between the identifier of an endpoint (i.e., what the endpoint is called) and its address (i.e., where the endpoint is located). Mechanisms for location management provide a time varying mapping between the identifier and the address of each endpoint. We have already introduced that location management typically consists of two operations: (1) updating (or registration), the process by which a mobile endpoint initiates a change in the location database according to its new location; and (2) finding (or paging), the process by which the network initiates a query for an endpoint's location (which may also result in an update to the location database). Most location management techniques use a combination of updating and paging in an effort to select the best trade-off between update overhead and latency involved in paging. Specifically, updates are not usually sent every time an endpoint enters a new cell, but rather are sent according to a predefined strategy such that the finding operation can be restricted to a specific area. There is also a trade-off, analyzed formally [1], between the update and paging costs. Location management methods as adapted in current cellular networks, such as GSM, Interim Standard 54 (IS-54), and IS-95, may be perceived as updating and querying a distributed database (the location register database) of endpoint identifier-to-address mappings. So it is important to determine when and how a change in a location register database entry should be initiated and how to organize and maintain the location register database.

4.1 Location Update

As MTs move around the network service area, the data stored in the location register databases may no longer be accurate. To ensure that calls can be delivered successfully, a mechanism is needed to update the databases with up-to-date location information. This is called location update (LU) or registration. There are several location updating methods based on LA structuring. Two automatic LA management schemes are very much in use [2]. These are (1) periodic location updating (although the simplest method, it has the inherent drawback of having excessive resource consumption which is unnecessary at times); and (2) location updating on LA crossing. Figure 4.1 illustrates a classification of possible update strategies [2].

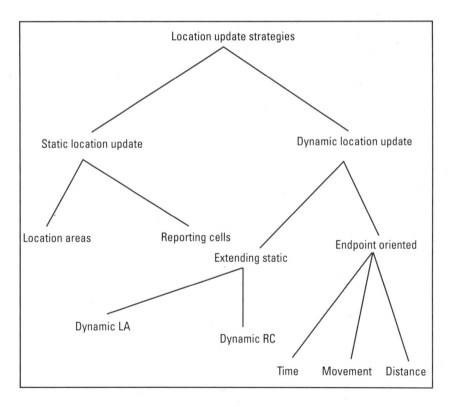

Figure 4.1 Classifications of location update strategies.

4.1.1   Location Update Static Strategies

In a static update strategy, there is a predetermined set of cells at which location updates may be generated. Whatever the nature of mobility of an endpoint, location updates may only be generated when, but not necessarily every time, the endpoint enters one of these cells. Two approaches to static updating are as follows:

1. *LAs (also referred to as paging or registration areas)* [3]: In this approach, the service area is partitioned into groups of cells with each group as an LA. An endpoint's position is updated if and only if the endpoint changes LAs. When an endpoint needs to be located, paging is done over the most recent LA visited by the endpoint. Location tracking in many second-generation cellular systems, including GSM [4] and IS-41 [5], is based on LAs [6]. Several strategies for LA planning in a city environment are evaluated in [7]. These include strategies that take into account geographical criteria (such as population distribution and highway topology) and user mobility characteristics.

2. *Reporting cells (or reporting centers)* [8]: In this approach, reporting cells, a subset of the cells, are designated as the only ones from which an endpoint's location may be updated. When an endpoint needs to be located, a search is conducted in the vicinity of the reporting cell from which the most recent update was generated. In [8], the problem of which cells should be designated as reporting cells in order to optimize a cost function is addressed for various cell topologies.

The principal drawback to static update strategies is that they do not accurately account for user mobility and frequency of incoming calls. For example, even though a mobile endpoint may remain within a small area, it may cause frequent location updates if that area happens to contain a reporting cell.

4.1.2   Location Update Dynamic Strategies

In a dynamic update strategy, an endpoint determines when an update should be generated, based on its movement. Thus, an update may be generated in any cell. A natural approach to dynamic strategies is to extend the static strategies to incorporate call and mobility patterns. The dynamic LA strategy proposed in [9] dynamically determines the size of an endpoint's LA according to the endpoint's incoming call arrival rate and mobility. Analytical results presented in [9] indicate that this strategy is an improvement over static strategies when call arrival rates are user-dependent or time-dependent. The dynamic reporting centers strategy proposed in [10] uses easily obtainable information to customize the choice of the next set of reporting cells at the time of each location update. In particular, the strategy uses information recorded at the time of the endpoint's last location update, including the direction of motion, to construct an asymmetric distance-based cell boundary and to optimize the cell search order. In [11], three dynamic strategies are described in which an endpoint generates a location update (1) every T seconds (time-based), (2) after every M cell crossings (movement-based), or (3) whenever the distance covered (in terms of number of cells) exceeds D (distance-based). Distance-based strategies are inherently the most difficult to implement since the mobile endpoints need information about the topology of the cellular network. It was shown in [11], however, that for memoryless movement patterns on a ring topology, distance-based updating outperforms both time-based and movement-based updating. In [1], a set of dynamic programming equations is derived and used to determine an optimal updating policy for each endpoint, and this optimal policy is in fact distance-based. Strategies that minimize location-tracking costs under specified delay constraints (i.e., the time required to locate an endpoint) have also been proposed. In [12], a paging procedure is described that minimizes the mean number of locations polled with a constraint on polling delay, given a probability distribution for endpoint locations. A distance-based update scheme and a complementary paging scheme that guarantee a predefined maximum delay on locating an endpoint are described in [13]. This scheme uses an iterative algorithm to determine the optimal update distance D that results in minimum cost within the delay bound.

In organizing the location database, one seeks to minimize both the latency and the overhead, in terms of the amount of storage and the number of messages required, in accessing location information. These are, in general, counteracting optimization criteria. Most solutions to the location database organization problem select a point, which is a three-way trade-off between overhead, latency, and simplicity. The simplest approach to location database organization is to store all endpoint identifier-to-address mappings in a single central place. For large numbers of reasonably mobile endpoints, however, this approach becomes infeasible in terms of database access time and storage space and also represents a single point of failure.

The next logical step in location database organization is to partition the network into a number of smaller pieces and place a portion of the location database in each piece. Such a distributed approach is well suited to systems where each subscriber is registered in a particular area or home. With this organization, the location database in an area contains the locations of all endpoints that have that area for a home. When the endpoint moves out of its home area, it updates its home location database to reflect the new location. The HLR and VLR schemes of emerging wireless cellular networks [6] are an example of this approach, as are the Mobile IP scheme [14] for the Internet and the GSM-based General Packet Radio Switching (GPRS) Network for data transport over cellular networks. Studies [15, 16] have shown that with predicted levels of mobile users, signaling traffic may exceed acceptable levels. Thus, researchers have considered augmenting this basic scheme to increase its efficiency under certain circumstances. For instance, in [17], per-user caching is used to reuse location information about a called user for subsequent calls to that user, and is particularly beneficial for users with high call-to-mobility ratios (i.e., the frequency of incoming calls is much larger than the frequency of location updates). In [18], "local anchoring" is used to reduce the message overhead by reporting location changes to a nearby VLR instead of to the HLR, thus increasing the location-tracking efficiency when the call-to-mobility ratio is low and the update cost is high. As with most large organizational problems, a hierarchical approach provides the most general and scalable solution. By hierarchically organizing the location database, one can exploit the fact that many movements are local. Specifically, by confining LU propagation to the lowest level (in the hierarchy) containing the moving endpoint, costs can be made proportional to the distance moved. Several works address this basic theme. In [19], a hierarchy of regional directories is prescribed where each regional directory is based on a decomposition of the network into regions. Here, the purpose of the ith-level regional directory is to enable tracking of any user residing within a distance of 2^i. This strategy guarantees overheads that are polylogarithmic in the size and diameter of the network. In [20], the location database is organized so as to minimize the total rate of accesses and updates. This approach takes into account estimates of mobility and calling rates between cells and a budget on access and update rates at each database site. In [21], location database organization takes into account the user profile of an endpoint (i.e., the predefined pattern of movement for the endpoint). Partitions of the location database are obtained by grouping the locations among which the endpoint moves frequently and by separating those to which the endpoint relocates infrequently. Each partition is further partitioned in a recursive fashion, along the same lines, to obtain a location database hierarchy.

In the above strategies, the emphasis is on reducing update overhead, but it is equally important to reduce database access latency. One strategy for doing so is replication, where identical copies of the database are kept in various parts of the network so that an endpoint location may be obtained using a low-latency query to a nearby server. The problem here is to decide where to store the replications. This is similar to the classical database allocation [22] and file allocation [23] problems, in which databases or files are replicated at sites based on query-update or read-write access patterns. In [24], the best zones for replication are chosen per endpoint location entry, using a minimum-cost maximum-flow algorithm to decide where to replicate the database, based on the calling and mobility patterns for that endpoint.

4.2 Paging

Paging involves messages sent over the radio informing the mobile user that an incoming call is pending. When the MS replies, the exact BS to which it is attached will be know to the network, and the call setup can proceed. The network knows the position of the MS only at the LA level. Since radio spectrum is scarce, these messages must be kept to a minimum by paging a minimum of cells. The trade-off, as pointed out above, is that in order to minimize the number of cells that must be paged, location updates must be more frequent. It should be taken into account that, because of the unpredictable nature of radio communications, paging messages may not arrive at the mobile with the first attempt, and there is usually some number of repetitions. Since the arrival of paging messages cannot be predicted, an MS must listen to the paging channel continuously or almost continuously as it is explained in GSM. In most operational systems, LA and PA are identical, or PAs are a subset of LA (see Figure 4.2). For this reason, any grouping of cells for location management purposes is usually called an LA.

4.2.1 Blanket Paging

Two major steps are involved in call delivery. These are (1) determining the serving VLR of the called MT, and (2) locating the visiting cell of the called MT. Locating the serving VLR of the MT involves the following database lookup procedures [25] (see Figure 4.3):

1. The calling MT sends a call initiation signal to the serving MSC of the MT through a nearby BS.

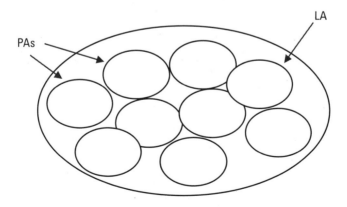

Figure 4.2 A number of PAs within LA.

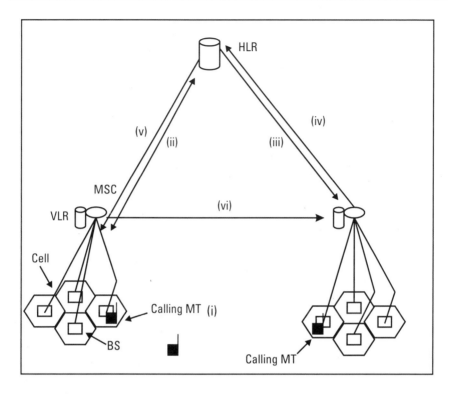

Figure 4.3 Call delivery procedure.

2. The MSC determines the address of the HLR of the called MT by global title translation (GTT) and sends a location request message to the HLR.

3. The HLR determines the serving VLR of the called MT and sends a route request message to the MSC serving the MT.

4. The MSC allocates a temporary identifier called temporary local directory number (TLDN) to the MT and sends a reply to the HLR together with the TLDN.

5. The HLR forwards this information to the MSC of the calling MT.

6. The calling MSC requests a call setup to the called MSC through CCS 7 network.

The procedure described above allows the network to set up a connection from the calling MT to the serving MSC of the called MT. Since each MSC is associated with an LA, a mechanism is therefore necessary to determine the cell location of the called MT. In current cellular networks, this is achieved by a paging procedure such that polling signals are broadcast to all the cells within the residing LA of the called MT over a forward control channel. On receiving the polling signal, the MT sends a reply over a backward control channel, which allows the MSC to determine its current residing cell. This is called blanket-paging method. In a selective paging scheme, instead of polling all the cells in an LA, a few cells are polled at a time. The cluster of cells polled at a time constitutes the PA. Here, a factor called the granularity factor, K is defined as the ratio of the number of cells in the PA to the number of cells in the LA. K denotes the fineness in the polling scheme. In a purely sequential polling scheme, $K = 1/S_j$ where the granularity factor is one in the case of blanket polling, and S_j is the number of cells in the jth LA.

4.2.2 Different Paging Procedures

Rose [26] developed methods for balancing call registration and paging. The probability distribution on the user location as a function of time is either known or can be calculated, the lower bounds on the average cost of paging are used in conjunction with a Poisson incoming-call arrival model to formulate the paging and registration optimization problem in terms of timeout parameters.

In their other work [12], efficient paging procedures are used to minimize the amount of bandwidth expended in locating a mobile unit. Given the probability distribution on user location, they have shown that the optimal paging strategy, which minimizes the expected number of locations polled, is to query each location sequentially in order of decreasing probability. Since a sequential search over many locations may impose unacceptable polling delay, they considered optimal paging subject to delay constraint.

In [13], they have proposed a mobile user location mechanism that incorporates a distance-based LU scheme and a paging mechanism that satisfied predefined delay requirements. In [27], they have introduced a mobility tracking mechanism that combines a movement-based LU policy with a selective paging scheme. This selective paging scheme decreases the location tracking cost under a small increase in the allowable paging delay.

In [28], they explored tracking strategies for MUs in personal communication networks, which are based on the topology of cells. The notion of topology-based strategies was introduced in a general form in this work. In particular, the known PAs, overlapping PAs, reporting centers and distance-based strategies were covered by this notion.

This work [29] proposed a method, which aims at the reduction of signaling overhead on the radio link, produced by the paging procedure. The key idea is the application of a multiple step paging strategy, which operates as follows: At the instance of a call terminating to an MU which roams within a certain LA, paging is initially performed in a portion of the LA (the PA) that the so-called "paging related information" indicates. Upon no paging response, the MU is paged in the complementary portion of the LA—this phase can be completed in more than one (paging) step. Several "intelligent" paging strategies were defined in this work. In [30], various paging schemes were presented for locating MUs in wireless networks. Paging costs and delay bounds are considered since paging costs are associated with bandwidth utilization and delay bounds influence call setup time. To reduce the paging costs, three paging schemes, reverse, semireverse, and uniform, were introduced to provide a simple way of partitioning the service areas and decrease the paging costs based on each MT's location probability distribution.

The several paging strategies mainly based on blanket paging were applied to reduce the paging costs as well as update costs associated with constraints. These strategies discussed briefly above were widely used and few of them are applied in industry. In spite of having widespread use of those paging strategies, some disadvantages have been found. In the next section, a few new paging schemes are discussed to overcome the disadvantages in the different blanket paging schemes.

4.3 Intelligent Paging Scheme

The movement of MTs is modeled according to some ergodic, stochastic process [31]. To provide ubiquitous communication link, irrespective of the location of MTs, the BSs provide a continuous coverage during the call and in the idle state also. When an incoming call comes to an MT, which roams within a certain LA, paging is initially performed within a portion of the LA, which is a subset of the actual LA. This portion of the LA, which is a set of base stations of paging (BSPs), is termed a PA. Intelligent paging is a multistep paging strategy [7], which aims at determining the proper PA within which the called MT currently roams. In order to quantitatively evaluate the average cost of paging, time-varying probability distributions on MTs are required. These distributions may be derived from the specific motion models, approximated via empirical data, or even provided by the MTs in the form of a partial itinerary at the time of the last contact. It is assumed that

1. The probability density function of the speed of MTs is known.
2. The process of movement of MTs is isotropic, Brownian motion [32] with drift. In the one-dimensional version of Brownian motion, an MT moves by one step Δx to the right with some probability, p and to the left with probability q, and stays there with probability $(1 - p - q)$ for each time step Δt. Given that the MT starts at time $t = 0$ for position $x = 0$, the Gaussian PDF on the location of an MT is given by:

$$P_{x(t)}(x(t)) = (\pi Dt)^{-0.5} e^{-k(x-vt)^*(x-vt)/Dt} \tag{4.1}$$

where $v = (p - q) \times (\Delta x/\Delta t)$ is the drift velocity and $D = 2((1 - p)\, p + (1 - q)\, q + 2pq)\, (\Delta x)^2/\, \Delta t$ is the diffusion constant, both functions of the relative values of time and space steps. Drift is defined as mean velocity in a given direction and is used to model directed traffic such as vehicles along a highway;

3. Time elapsed since the last known location;

4. The paging process described here is rapid enough for the rate of motion of the MT (i.e., the MT to be found does not change its location during the paging process).

The algorithm of an intelligent paging process on arrival of a paging request (PR) is given below:

```
While PR is attached {
   while MT is not busy {
      if current traffic load exceeds threshold traffic
      load {
         initialize the incremental counter i = 0;
      select the proper PA;
      page within the selected PA;
      if reply against PR received
         then stop;
      else {
         while (imaximum value of incremental counter i)
            do {
               page within another PA;
               increase the incremental counter i = i + 1;
               }
            }
         }
      else
         apply blanket paging;
   }
}
```

The intelligent paging strategy maps the cells inside the LA comprising S cells into a probability line at the time of arrival of the incoming call. This mapping depends on factors such as the mobility of the MT, its speed profile, the incoming call statistics, and the state of the MT at that instant. This procedure is called attachment. If it is detached, the PR is cancelled. If it is busy, a relation between the MT and the network already exists, and therefore paging is not required. If it is free, the network proceeds for paging upon receipt of a PR (see above algorithm). In the intelligent paging scheme, the network determines the probability of occupancy of the called MT in different cells in an LA. These probabilities are arranged in descending order. The order in which cells are to be polled depends on the ordered set of probability occupancy vector. In each paging cycle, the MSC serves PRs stored in its buffer, independently of each other. It is done by assigning a BS to each of the n requests in the buffer according to an assignment policy. The jth PR may be sent to the ith BS, in order to be paged in the corresponding cell. There are many ways to generate such assignments. Two methods will be presented subsequently in this chapter as part of the proposed intelligent paging scheme. If the buffer size is n and there are k PAs (denoted by $A_1, A_2,..., A_k$), then we can write:

$$A_1 \cup A_2 \cup ..... A_k = S$$

$$Ai \cup Aj = \phi$$

This means that the PAs are mutually exclusive and collectively exhaustive. Paging and channel allocation packets from a BS to MTs are multiplexed stream in a forward signaling channel. Paging rate represents the average number of paging packets, which arrives at a BS during unit time. Paging signals are sent to the BSs via landlines and are broadcast over the forward signaling channel. As each attached MT in the LA constantly monitors the paging channels to check whether it is paged or not, the distributor in the MSC which is a part of Mobility Manager (MM) allocates the distribution of PRs to the BSs for each paging cycle based on the information collected over the previous paging cycle. As soon as an MT is found, the corresponding PR is purged from the buffer and a new PR replaces it. The function of the distributor in the MM is to map the PRs to the PAs:

$$g : (PR_1, PR_2,..., PR_n) \rightarrow (A_1, A_2,..., A_k)$$

Depending on the nature of polling, there may be two types of search. These are sequential and parallel-o-sequential. In purely sequential polling, one cell is polled in a paging cycle. Sometimes, due to delay constraint, instead of polling one cell at a time, we poll a cluster of cells in an LA. This is called parallel-o-sequential intelligent paging (PSIP) which is a special case of sequential intelligent polling (SIP).

The network first examines whether a multiple-step paging strategy should be applied or not. The decision is based on the current traffic load. Normally, when this load exceeds a threshold value, a multiple-step paging strategy is employed. In the very first phase, the network decides whether checking the current status of the MT needs paging. The network then examines whether the appropriate type of paging is blanket paging or multiple-step paging. The granularity factor K shows fineness in polling. In general, the granularity factor is defined as

K = (number of cells to be polled in a cycle)/(number of cells in an LA)

The maximum value of the granularity factor is 1, when all cells in an LA are polled in one polling cycle. The granularity factor in SIP is, K^{SIP} = 1/(number of cells in an LA). The granularity factor in PSIP is K^{PSIP} = (number of cells in the cluster)/(number of cells in an LA).

We assumed a perfect paging mechanism where an MT will always respond to a paging signal meant for it, provided it receives the PR. However, situations may arise that leave an MT undetected even though the distributor in the MSC is able to select the corresponding BS and initiate a PR for it. Such a situation will arise when there are more PRs assigned to a BS by the distributor than the number of paging channels available in a cycle. As there are only l paging channels per BS, the PRs in excess of l will be considered to be blocked. These excess PRs will be attempted for sending to select BSs in subsequent paging cycles. So the called MT may be inside the area of an overloaded cell. But the PR for it might be blocked in a paging cycle. So, the distributor must keep track of the number of times a search has been attempted for the PR.

The application of intelligent paging includes the event of paging failures due to wrong predictions of the locations of the called MT. In such cases, another step, or more than one step, will be required (i.e., the called MT will be paged in other PAs). Continuous unsuccessful paging attempts may lead to unacceptable network performance in terms of paging delay. Moreover, the paging cost will also increase with each unsuccessful attempt to locate the called MT. In such cases, the network does not preclude the option of single-step paging at a certain intermediate point of search.

4.3.1 Sequential Intelligent Paging

In sequential intelligent paging schemes, one cell is polled at a time and the process continues until such time the called MT is found or timeout occurs whichever is earlier. The selection of the cell to be polled sequentially depends on the value of the occupancy probability vector, which is based on the stochastic modeling delineating the movement of the MT. In SIP, the PRs are stored in a buffer of the MSC and each PR is sent to that BS where there is maximum probability of finding the called MT. When the paging is unsuccessful during a polling cycle the MT is sequentially paged in other cells of the LA that have not been polled so far. This phase is completed in one, or more than one, paging step(s). The sequential paging algorithm is given below:

```
STEP 1:      When an incoming call arrives, calculate the
             occupancy probability vector, [P] of an MT for
             the cells in the LA based on the probability
             density function, which characterizes the
             motion of the MT;
STEP 2:      Sort the elements of [P] in descending order;
STEP 3.0:    FLAG = False;
             i = 1;
STEP. 3.1:   Poll the ith cell for i(S;
STEP. 3.2:   If the MT is found
             FLAG = True;
             Go to ENDSTEP;
STEP. 4.0:
             If timeout occurs
               Go to ENDSTEP;
             Else
               i = i + 1;
               Go to STEP 3.1;
             Endif
ENDSTEP:     If FLAG = True
               "Declare  Polling is Successful";
             Else
             "Declare    Polling is Unsuccessful";
             Endif
```

As extra processing is required at the MSC, an inherent delay will be associated with this process (i.e., before the PR is sent to the appropriate BS). This delay includes the determination of the probabilities in different cells, sorting of these probabilities in descending order and polling the cells sequentially depending on those values. This delay will be added to the call setup process. The amount of this delay will be $\sim (O\,(S) + O\,(S \log S) + O\,(S/2))$.

4.3.2 PSIP

PSIP is a special case of SIP where K 1. Instead of polling a single cell in each cycle, here we partition the LAs into several PAs and poll those PAs sequentially comprising more than one cell. The benefit that accrues out of PSIP is a significant improvement in the expected discovery rate of called MTs and the overwhelming reduction in paging cost and signaling load. The number of steps in which the paging process should be completed depends on the allowed delay during paging. The application of PSIP also includes the event of paging failures due to unsuccessful predictions of the location of called MT. In such cases, multiple steps are required and the called MT is paged in another portion of the LA. To obviate the deterioration of the network performance in such a situation and minimize the number of paging steps, the network should guarantee the formation of PAs such that the P_{SFP} is high (typical value >90%). So, the PA should consist of those cells where the sum of probabilities of finding the called MT is greater than or equal to the typical value chosen for P_{SFP}. The parallel-o-sequential paging algorithm is given below:

```
STEP 1:   When an incoming call arrives, find out the cur
          rent state of the called MT;
STEP 2:   If MT is detached
              PR is cancelled;
              Go to ENDSTEP;
     Else
              If MT is busy (location is known)
                 go to ENDSTEP;
     Else
              Find granularity factor K;
     Endif
STEP 3:   If granularity factor is 1
              Poll all the cells;
              go to ENDSTEP;
     Else
              find out [P] , the occupancy probability vector
              of the MT for the cells in the LA, based on the
              probability density function which characterizes
              the motion of the MT;
     Endif
STEP 4:   Sort the elements of [P] in descending order;
          Set all the cells as  unmarked ;
STEP 5.0: Select a proper PA consisting of "unmarked"
          cells for which Σpi>P_SFP;
          FLAG = False;
STEP 5.1: Poll ith cluster and label the cells in ith
          cluster as  marked
STEP 5.2: if the MT is found
          FLAG = True;
          Go to ENDSTEP;
STEP 6.0:
              If timeout occurs
              Go to ENDSTEP;
     Else
          Go to STEP 5.0;
     Endif
ENDSTEP:  If FLAG = True
          Declare:  "Polling is Successful" ;
     Else
          Declare:   "Polling is Unsuccessful" ;
     Endif
```

4.3.3   Comparison of Paging Costs

In the conventional or the blanket paging, upon arrival of an incoming call, the paging message is broadcast from all the BSs in the LA. That means all the cells in the LA are polled at a time for locating the called MT (i.e., each MT is paged S times before the called MT is discovered). The polling cost per cycle is

$$C_p^{\text{conv}} = SA_{\text{cell}}\rho\mu T_p B_p \tag{4.2}$$

The SIP strategy, described here, aims at the significant reduction in load of paging signaling on the radio link by paging a cell sequentially. PRs arrive according to a Poisson process at the buffer of the MSC. The distributor issues the PRs to appropriate BSs. These PRs are queued at the location and serviced on an FCFS basis at the average rate ζ. The result may be a success or a failure. The results of completed polls are fed back to the controller in the BS for further appropriate action. As pointed out earlier, a called MT may not be found during a paging cycle. Either the number of paging channels may be insufficient to accommodate the PR in a particular cycle, or the search for the called MT in the cell results in a failure. In both cases, the polling process goes through more than one cycle. So, the paging cost per polling cycle in this scheme is

$$C_p{}^{SIP} = K^{SIP} SA_{cell} \rho\mu(1+z)T_p B_p \qquad (4.3)$$

The variable z accounts for the unsuccessful PRs from the previous cycle due to either of the two reasons, so z depends on the success rate, timeout duration, and the number of paging channels available per BS. In GSM, assuming that a sufficient channel for paging is there, z becomes zero. In the best case (i.e., when the called MT is found during the first polling cycle of SIP), z is also zero. In the worst case, all the cells in the LA are to be polled before the MT is found. Then the polling cost just exceeds that in GSM. Moreover, the delay is also at a maximum (i.e., S units). There may be a situation when the polling cost in a SIP scheme exceeds the cost in blanket polling significantly. If the MT resides in a cell with a low probability of occupancy and returns to one of the cells that is polled already after the polling cycle, the called MT will not be found even after polling all the cells in the LA. Such an incidence is likely when the number of cells is greater, a few cells have the same probability of occupancy, and the MT is very mobile. In this case, the call is blocked or the cells are polled sequentially once again to find the MT. So, K has an inverse effect on z. The granularity factor K is generally chosen to be more than one to avoid such a scenario. The paging cost per polling cycle in PSIP is

$$C_p{}^{SIP} = K^{SIP} SA_{cell} \rho\mu(1+z)T_p B_p \qquad (4.4)$$

The variable z also accounts for the unsuccessful PRs from the previous cycle. As mentioned earlier also, K is chosen such that $P_{SFP} > 0.9$. The optimum value of K varies from case to case.

4.4 More Paging Schemes

Assume that each LA consists of the same number N of cells in the system [30]. The worst-case paging delay is considered as delay bound D in terms of the polling cycle. When D is equal to 1, the system should find the called MT in one polling cycle, requiring all cells within the LA to be polled simultaneously. The paging cost C, which is the number of cells polled to find the called MT, is equal to N. In this case, the average paging delay is at its minimum, which is one polling cycle, and the average paging cost is at its maximum, $C = N$. On the other hand, when D is equal to N, the system will poll one cell in each polling cycle and search all cells one by one. Thus, the worst-case occurs when the called MT is found in the last polling cycle, which means the paging delay would be at its maximum and equal to N polling cycles [33]. The average paging cost, however, may be minimized if the cells are searched in decreasing order of location probabilities [12].

Consider the partition of PAs given that $1 \le D \le N$, which requires grouping cells within an LA into the smaller PAs under delay bound D. Suppose, at a given time, the initial state P is defined as P= $[p_1, p_2,..., p_j,..., p_N]$, where p_j is the location probability of the jth cell to be searched in decreasing order of probability. Thus, the time effect is reflected in the location probability distribution. We use triplets $PA^*_p (i, q_i, n_i)$ to denote the PAs under the paging scheme P in which i is the sequence number of the PA, q_i is the location probability that the called MT can be found within the ith PA, and n_i is the number of cells contained in this PA. In Figure 4.4, an LA is divided into D PAs because the delay bound is assumed to be D. Thus, the worst-case delay is guaranteed to be D polling cycles. The system searches the PAs one after another until the called MT is found. Three paging schemes are discussed in this section.

4.4.1 Reverse Paging

This scheme is designed for a situation where the called MT is most likely to be found within a few cells. Consider the first $(D - 1)$ highest probability cells as the first $(D - 1)$ PAs to be searched. Each of these $(D - 1)$ PAs consists of only

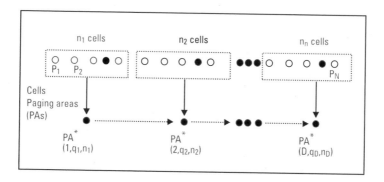

Figure 4.4 Partition of LA into PAs.

one cell. Then the remaining $(N - D + 1)$ lower probability cells are lumped together to be the last PA, (i.e., the Dth PA). The newly formed PAs become $\text{PA}^*_r (1, p_1, 1), \text{PA}^*_r (2, p_2, 1),..., \text{PA}^*_r (D-1, p_{D-1}, 1), \text{PA}^*_r (D, q_D, N-D+1)$, where r denotes the reverse paging scheme.

4.4.2   Semireverse Paging

Since the average paging cost can be minimized by searching cells in decreasing order of location probability, if a delay bound D is not applied [12], intuitively, the paging cost can be reduced by searching the PAs in decreasing order of probability. Under semireverse paging schemes, a set of PAs is created in a nonincreasing order of location probabilities. First, combine the two cells with the lowest location probabilities into one PA, and then reorder all PAs in nonincreasing order of location probabilities. Keep combining the two PAs with the lowest probabilities and reorder them until the total number of PAs is equal to D. If two PAs have the same probability, the PA with fewer cells has higher priority, (i.e., its sequence number is smaller). Semireverse paging schemes guarantee that the location probability of each PA is in a nonincreasing order. However, the cell with lower probability may be searched before the cell with higher probability because the initial sequence of the cells is reordered during the semireverse paging procedure.

4.4.3   Uniform Paging

Under a uniform paging scheme, the LA is partitioned into a series of PAs in such a way that all PAs consist of approximately the same number of cells. The *uniform paging* procedure is as follows:

- Calculate the number of cells in each PA as $n_0 = \lfloor N/D \rfloor$ where N = $n_0 D + k$.

- Determine a series of PAs as $\text{PA}^*_u (1, p_1, 1), \text{PA}^*_u (2, p_2, 1),..., \text{PA}^*_u (D, q_D, n_D)$. Note that there are n_0 cells in each of the first $(D - k)$ PAs and there are $n_0 + 1$ cells in each of the remaining PAs. This means $n_1 = n_2 =... = n_{D-k} = n_0$, and $n_{D-k+1} =... = n_D = n_0 + 1$. For example, the first PA consists of n_0 cells and the last PA, (i.e., Dth PA), consists of $n_0 + 1$ cells.

- The network polls one PA after another sequentially until the called MT is found.

4.5 Intersystem Paging

In a multitier wireless service area consisting of dissimilar systems, it is desirable to consider some factors, which will influence the radio connections of the MTs roaming between different systems [34]. As shown in Figure 4.5, there are two systems, *Y* and *W* in the microcell tier, which may use different protocols such as DCS1800 and PCS1900. Each hexagon represents an LA within a stand-alone system, and each LA is composed of a cluster of microcells. The terminals are required to update their location information with the system whenever they enter a new LA; therefore, the system knows the residing LA of a terminal all the time. In the macrocell tier there are also two systems, *X* and *Z* in which different protocols (e.g., GSM and IS-41) are applied. For macrocell systems, one LA can be one macrocell. It is possible that systems *X* and *W*, although in different tiers, may employ similar protocols such as IS-95, GSM, or any other protocol. There are two types of roaming shown in Figure 4.5: intrasystem and intersystem roaming. Intrasystem roaming refers to an MT's movement between the LAs within a system such as *Y* and *Z*. Intersystem roaming refers to the MTs that move between different systems. For example, mobile users may travel from a macrocell system within an IS-41 network to a region that uses GSM standard.

For intersystem location update, a boundary region called boundary LA (BLA) exists at the boundary between two systems in different tiers [34]. In addition to the concept of BLA, a boundary location register (BLR) is embedded

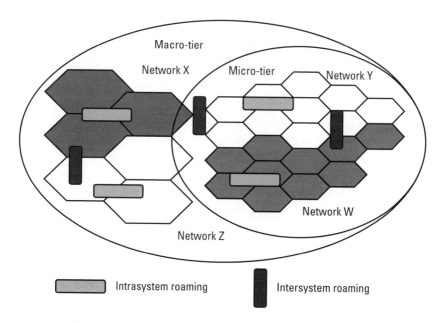

Figure 4.5 Architecture of intrasystem and intersystem roaming.

in the boundry interworking unit (BIU). A BLR is a database cache to maintain the roaming information of MTs moving between different systems. The roaming information is captured when the MT requests a location registration in the BLA. The BLRs enable the intersystem paging to be implemented within the appropriate system that an MT is currently residing in, thus reducing the paging costs. Therefore, the BLR and the BIU are accessible to the two adjacent systems and are collocated to handle the intersystem roaming of MTs. The VLR and the MSC, however, provide roaming information within a system and deal with the intrasystem roaming of MTs. Besides, there is only one BLR and one BIU between a pair of neighboring systems, but there may be many VLRs and MSCs within a stand-alone system.

When a call connection request arrives at X, the call will be routed to the last registered LA of the called MT. Given that the last registered LA within X is adjacent to Y, the system needs to perform the following steps to locate the MT:

1. Sends a query signal to the BLR between X and Y to obtain the MT's location information. This step is used to ascertain whether the MT has crossed the boundary or not.

2. If the MT has already moved to Y, searches only the LA in Y. Otherwise, the last registered LA within X will be searched. Within network X or Y, one or multiple polling messages are sent to the cells in the LA according to a specific paging scheme.

As a result, only one system (X or Y) is searched in the paging process for intersystem roaming terminals. This approach will significantly reduce the signaling cost caused by intersystem paging. In particular, it is very suitable for the high traffic environment because it omits the searching of two adjacent systems. Moreover, since the BLR is an additional level of cache database, it will not affect the original database architecture. Another advantage of the BLR is that it reduces the zigzag effect caused by intersystem roaming. For example, when an MT is moving back and forth on the boundary, it only needs to update the information in the BLR instead of contacting the HLRs. If the new BLR concept is not used, the intersystem paging can still take place. The system will search X first, if the called MT cannot be found, then Y will be searched. This method increases the paging cost as well as the paging delay, thus degrading the system performance.

4.6   IP Micromobility and Paging

Recent research [35] in Mobile IP has proposed that IP should take support from the underlying wireless network architecture to achieve good performance for handover and paging protocols. Recent IETF work defines requirements for layer 2 (the data link layer of the OSI model) to support optimized layer 3 (the network layer of the OSI model) handover and paging protocols. Layer 2 can send notification to layer 3 that a certain event has happened or is about to happen. The notification is sent using a trigger. IP micromobility and paging protocols require support from underlying wireless infrastructure.

Layer 2 can help layer 3 by providing information in the form of triggers. A trigger can be defined as an abstraction of notification from layer 2 that a certain event has happened or is about to happen. In [36], various ways of implementing triggers are discussed. A trigger may be implemented using system calls. The operating system may allow an application thread to register callback for a layer 2 trigger, using system calls of an application-programming interface (API). A system call returns when that particular event is fired in layer 2. Each trigger is defined by three parameters:

1. The event that causes trigger to fire;
2. The entity that receives the trigger;
3. The parameters delivered with the trigger.

Triggers were defined to aid low-latency handover in Mobile IP [37]. Another set of triggers was defined in [38] to aid IP paging protocols. CDMA, for example, works in conjunction with Mobile IP to support mobility in IP hosts. IP paging is a protocol used to determine the location of a dormant (a mode that conserves battery by not performing frequent updates) Mobile Node (MN). Paging triggers were defined in [38] to aid movement of a MN in multiple IP subnets in the same layer 2 PA. Paging triggers aid the MN in entering dormant mode in a graceful manner and make the best use of paging provided by underlying wireless architecture.

4.7 Location Management

Location management schemes are essentially based on users' mobility and incoming call rate characteristics. It is two-stage process that enables the network to discover the current attachment point of the MU for call delivery. The first stage is location registration or location update. In this stage, the MT periodically notifies the network of it new access point, allowing the network to authenticate the user and revise the user's location profile. The second stage is call delivery. Here, the network is queried for the user location profile and the current position of the mobile host is found. Current techniques for location management involve database architecture design and the transmission of signaling messages between various components of a signaling network. Other issues include: security, dynamic database updates, querying delays, terminal paging methods, and paging delays.

There are two standards for location management currently available: Electronics and Telecommunications Industry Associations (EIA/TIA) IS-41 [39] and the GSM mobile application part (MAP) [6]. The IS-41 scheme is commonly used in North America for AMPS, IS-54, IS-136, and PACS networks, while the GSM MAP is mostly used in Europe for GSM and DCS1800 networks. Both standards are based on a two-level data hierarchy. Location registration procedures update the location databases (HLR and VLRs) and authenticate the MT when up-to-date location information of an MT is available. Call delivery procedures locate the MT based on the information available at the HLR and VLRs when a call for an MT is initiated. The IS-41 and GSM MAP location management strategies are very similar. While GSM MAP is designed to facilitate personal mobility and to enable user selection of network provider, there are a lot of commonalities between the two standards [4, 25]. The location management scheme may be categorized in several ways.

4.7.1   Without Location Management

In this *level 0* method, no location management is realized [40]; the system does not track the mobiles. A search for a called user must therefore be done over the complete radio coverage area and within a limited time. This method is usually referred to as the flooding algorithm [41]. It is used in paging systems because of the lack of an uplink channel allowing a mobile to inform the network of its whereabouts. It is also used in small private mobile networks because of their small coverage areas and user populations. The main advantage of not locating the MTs is obviously simplicity; in particular, there is no need to implement special databases. Unfortunately, it does not fit large networks dealing with high numbers of users and high incoming call rates.

4.7.2   Manual Registration in Location Management

This *level 1* method [40] is relatively simple to manage because it only requires the management of an indicator, which stores the current location of the user. The mobile is also relatively simple; its task is just limited to scanning the channels to detect paging messages. This method is currently used in telepoint cordless systems (such as CT2). The user has to register each time he or she moves to a new island of CT2 beacons. To page a user, the network first transmits messages through the beacon with which he or she registered and, if the mobile does not answer, extends the paging to neighboring beacons. The main drawback of this method is the constraint for a user to register each time he or she moves.

4.7.3   Automatic Location Management Using LA

Presently, this *level 2* location method [40] most widely implemented in first- and second-generation cellular systems (e.g., NMT, GSM, IS-95) makes use of LAs (see Figure 4.1). In these wide-area radio networks, location management is done automatically. LAs allow the system to track the mobiles during their roaming in the network(s): subscriber location is known if the system knows the LA in which the subscriber is located. When the system must establish a communication with the mobile (to route an incoming call, typically), the paging only occurs in the current user LA. Thus, resource consumption is limited to this LA; paging messages are only transmitted in the cells of this particular LA. Implementing LA-based methods requires the use of databases. Generally, a home database and several visitor databases are included in the network architecture.

4.7.4   Memoryless-Based Location Management Methods

All methods included are based on algorithms and network architecture, mainly on the processing capabilities of the system.

4.7.4.1   Database Architecture

LA partitioning, and thus mobility management cost, partly relies on the system architecture (e.g., database locations). Thus, designing an appropriate database organization can reduce signaling traffic. The various database architectures are proposed with this aim [25, 42–44]. An architecture where a unique centralized database is used is well-suited to small and medium networks, typically based on a star topology. The second one is a distributed database architecture, which uses several independent databases according to geographical proximity or service providers. It is best-suited to large networks including subnetworks managed by different operators and service providers. The GSM worldwide network, defined as the network made up of all interconnected GSM networks in the world, can be such an example of a large network. The third case is the hybrid database architecture that combines the centralized and distributed database architectures. In this case, a central database (HLR-like) is used to store all user information. Other smaller databases (VLR-like) are distributed all over the network. These VLR databases store portions of HLR user records. A single GSM network is an example of such architecture.

4.7.4.2   Optimizing Fixed Network Architecture

In second-generation cellular networks and third-generation systems, the intelligent network manages signaling [45]. Appropriately organizing mobility functions and entities can help reduce the signaling burden at the network side. The main advantage of these propositions is that they allow us to reduce the network mobility costs independent of the radio interface and LA organization.

4.7.4.3   Combining LAs and PAs

In current systems, an LA is defined as both an area in which to locate a user and an area in which to page him or her. LA size optimization is therefore achieved by taking into account two antagonistic procedures, locating and paging. Based on this observation, several proposals have defined location management procedures, which make use of LAs and PAs of different sizes [46]. One method often considered consists of splitting an LA into several PAs. An MS registers only once, that is, when it enters the LA. It does not register when moving between the different PAs of the same LA. For an incoming call, paging messages will be broadcast in the PAs according to a sequence determined by different strategies. For example, the first PA of the sequence can be the one where the MS was last detected by the network. The drawback of this method is the possible delay increase due to large LAs.

4.7.4.4   Multilayer LAs

In present location management methods, LU traffic is mainly concentrated in the cells of the LA border. Based on this observation and to overcome this problem, Okasaka has introduced the multiplayer concept [47]. In his method, each MS is assigned to a given group, and each group is assigned one or several layers of LAs. This location updating method, although it may help reduce channel congestion, does not help reduce the overall signaling load generated by LUs.

4.7.5   Memory-Based Location Management Methods

The design of memory-based location management methods has been motivated by the fact that systems perform many repetitive actions, which can be avoided if predicted. This is particularly the case for LUs. Indeed, present cellular systems achieve every day, at the same peak hours, almost the same LU processing. Systems act as memoryless processes.

4.7.5.1   Dynamic LA and PA Size

The size of LAs is optimized according to mean parameter values, which in practical situations vary over a wide range during the day and from one user to another. Based on this observation, it is proposed to manage user location by defining multilevel LAs in a hierarchical cellular structure [48]. At each level the LA size is different, and a cell belongs to different LAs of different sizes. According to past and present MS mobility behavior, the scheme dynamically changes the hierarchical level of the LA to which the MS registers. LU savings can thus be obtained.

An opposite approach considers that instead of defining LA sizes a priori, these can be adjusted dynamically for every user according to his or her incoming call rate a and LU rate u_k, for instance. In [9], a mobility cost function denoted $C(k, a, u_k)$ is minimized so that k is permanently adjusted. Each user is therefore related to a unique LA for which size k is adjusted according to his or her particular mobility and incoming call rate characteristics. Adapting the LA size to each user's parameter values may be difficult to manage in practical situations. This led, in [49], to the definition of a method where the LAs sizes are dynamically adjusted for the whole population, not per user.

4.7.5.2 Individual User Patterns

Observing that users show repetitive mobility patterns, the alternative strategy (AS) is defined [50, 51]; its main goal is to reduce the traffic related to mobility management—and thus reduce the LUs—by taking advantage of users' highly predictable patterns. In AS, the system handles a profile recording the most probable mobility patterns of each user. The profile of the user can be provided and updated manually by the subscriber himself or determined automatically by monitoring the subscriber's movements over a period of time. For an individual user, each period of time corresponds to a set of LA, k. When the user receives a call, the system pages him sequentially over the LA a_i until getting an acknowledgment from the mobile. When the subscriber moves away from the recorded zone $\{a_1,..., a_k\}$, the terminal processes a voluntary registration by pointing out its new LA to the network. The main savings allowed by this method are due to the nontriggered LUs when the user keeps moving inside his profile LAs. So, the more predictable the user's mobility, the lower the mobility management cost. The main advantage of this method relies on the reduction of LUs when a mobile goes back and forth between two LAs.

4.7.6   Location Management in Next-Generation Systems

The next generation in mobility management will enable different mobile networks to interoperate with each other to ensure terminal and personal mobility and global portability of network services. However, in order to ensure global mobility, the deployment and integration of both wire and wireless components are necessary. The examples given focus on issues related to mobility management in a future mobile communications system, in a scenario where different access networks are integrated into an IP core network by exploiting the principles of Mobile IP. Mobile IP [52], the current standard for IP-based mobility management, needs to be enhanced to meet the needs of future 4G cellular environments. In particular, the absence of a location management hierarchy leads to concerns about signaling scalability and handoff latency, especially for a future infrastructure that must provide global mobility support to potentially billions of mobile nodes and accommodate the stringent performance bounds associated with real-time multimedia traffic. In this book, the discussion is confined to Mobile IP to describe the aspects of location management in 3G systems and beyond.

The aim of 4G cellular networks is to develop a framework for truly ubiquitous IP-based access by mobile users, with special emphasis on the ability to use a wide variety of wireless and wired access technologies to access the common information infrastructure. While the 3G initiatives are almost exclusively directed at defining wide-area packet-based cellular technologies, the 4G vision embraces additional local-area access technologies, such as IEEE 802.11-based WLANs and Bluetooth-based wireless personal area networks (WPANs). The development of MTs with multiple physical or software-defined interfaces is expected to allow users to seamlessly switch between different access technologies, often with overlapping areas of coverage and dramatically different cell sizes.

Figure 4.6 shows one example [53] of this multitechnology vision at work in a corporate campus located in an urban environment. While conventional wide-area cellular coverage is available in all outdoor locations, the corporation

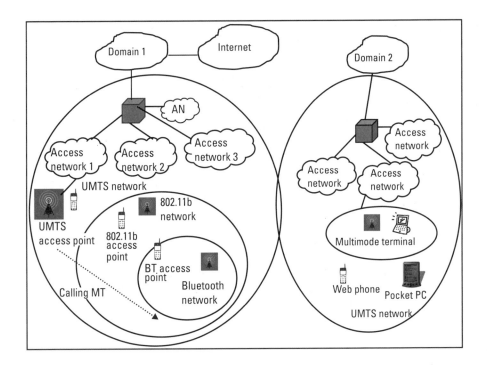

Figure 4.6 Heterogeneous network environments.

also offers 802.11-based access in public indoor locations such as the cafeteria and parking lots, as well as Bluetooth-based access to the Internet in every individual office. As MUs drive to work, their ongoing Voice over IP (VoIP) calls are seamlessly switched, first from the wide-area cellular to the WLAN infrastructure, and subsequently from the 802.11 access point to the Bluetooth access point located in their individual cubicles or offices. Since a domain can comprise multiple access technologies, mobility management protocols should be capable of handling vertical handoffs (i.e., handoffs between heterogeneous technologies).

Due to the different types of architecture envisaged in the multiaccess system, three levels of location management procedures can be envisaged [54]:

- *Internet (interdomain) network location management:* identifies the point of access to the Internet network;

- *Intrasegment location management:* is executed by segment-specific procedures when the terminal moves within the same access network;

- *Intersegment location management:* is executed by system-specific entities when the terminal moves from one access network to another.

In Mobile IP (see Figure 4.7) [52], each mobile node is assigned a pair of addresses. The first address is used for identification, known as the home IP address, which is defined in the address space of the home subnetwork. The second address is used to determine the current position of the node and is known as the care of address (CoA), which is defined in the address space of the visited or foreign subnetwork. The continuous tracking of the subscriber's CoA allows the Internet to provide subscribers with roaming services. The location of the subscriber is stored in a database, known as a binding table, in the home agent and in the corresponding node. By using the binding table, it is possible to route the IP packets toward the Internet point of access to which the subscriber is connected.

The terminal can be seen from the Internet perspective as an MT. Once the MT selects an access segment, the access point to the Internet network is automatically defined. The MT is therefore identified by a home address of the home subnetwork and by a CoA of the access segment. In the target system, location management in the Internet network is based on the main features of Mobile IP. Nevertheless, a major difference can be identified between the use of Mobile IP in fixed and mobile networks. In the fixed Internet network, IP packets are routed directly to the mobile node, whereas in the integrated system considered in this article, packets will be routed up to the appropriate edge router. Once a packet leaves the edge router and reaches the access network, the routing toward the final destination will be performed according to the mechanisms

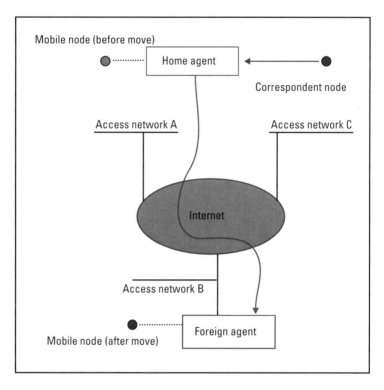

Figure 4.7 Mobile IP architecture.

adopted by each access segment (intrasegment mobility). When the MT decides to change access segments, its CoA will be changed. Therefore, the new CoA has to be stored in the corresponding binding tables. Since these binding tables can be seen as a type of location management database, this binding update can also be seen as a form of location update in the Internet.

Intersegment location management is used to store information on the access segments at a particular time. The information is then used to perform system registration, location update, and handover procedures. Using certain parameters, including the condition of the radio coverage, and QoS perceived by the user, the MT continuously executes procedures with the objective of selecting the most suitable access segment. Any modifications to these parameters could therefore lead to a change of access segment. This also implies a change in the point of access to the Internet network. Therefore, in order to route these packets correctly, it is necessary to have information on the active access segment, particularly information concerning the edge router that is connected to the node of the access segment. Thus, from the Internet point of view, no additional procedure or database is required since the information is implicitly contained in the CoA assigned to the MT.

4.8   LA Planning

LA planning in minimum cost plays an important role in cellular networks because of the trade-off caused by paging and registration signaling. The upper bound on the size of an LA is the service area of an MSC. In that extreme case, the cost of paging is at its maximum, but no registration is needed. On the other hand, if each cell is an LA, the paging cost is minimal, but the registration cost is at a maximum. In general, the most important component of these costs is the load on the signaling resources. Between the extremes lie one or more partitions of the MSC service area that minimize the total cost of paging and registration. In this section, a few approaches are discussed that address issues related to LA planning.

4.8.1   Two-Step Approach

The two-step approach [55] deals with the planning of LAs in a personal communication services network (PCSN) to be overlaid on an existing wired network. Given the average speed of MTs, the number of MSCs, their locations, the call handling capacity of each MSC, the handoff cost between adjacent cells, and the call arrival rate, an important consideration in a PCSN is to identify the cells in every LA to be connected to the corresponding MSC in a cost effective manner. While planning an LA, a two-step approach is presented, namely optimization of total system recurring cost (subproblem 1), and optimization of hybrid cost (subproblem 2). The planning first determines the optimum number of cells in an LA from subproblem 1. Then it finds out the exact LAs by assigning cells to the switches while optimizing the hybrid cost, which comprises the handoff cost and the cable cost, in subproblem 2. The decomposition of the problem provides a practical way for designing LAs. As this approach toward LA planning takes into accounts both cost and network planning factors, this unique combination is of great interest to PCSN designers. It develops an optimum network planning method for a wide range of call-to-mobility ratios (CMRs) that minimizes the total system recurring cost while still ensuring a good system performance. Approximate optimal results, with respect to cell-to-switch assignment, are achievable with a reasonable computational effort that supports the engineered plan of an existing PCSN.

4.8.1.1   Proposed Architecture

A multilevel hierarchical structure of PCSN, as shown in Figure 4.8, is considered in this work. At level 2, the total service area (SA) is divided into a number of LAs. Each LA is further subdivided into a number of cells at level 1. For each cell, there is a BS to provide the radio interface to the MTs within the cell. The BSs within an LA talk with each other through an MSC, situated at level 2,

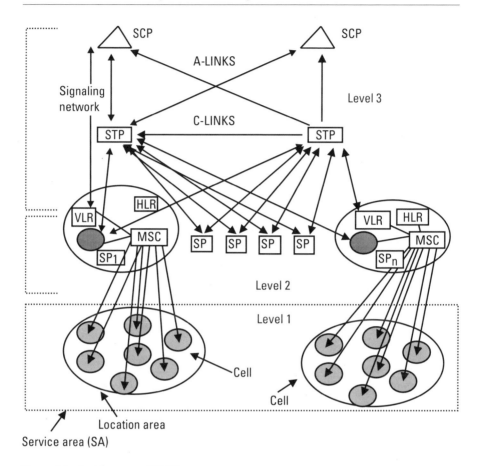

Figure 4.8 Architecture of PCSN.

for that LA. Thus, Each LA is serviced by an MSC, which is collocated with the signaling points (SPs) of a CCS-7 network [56]. The SPs are interconnected by C-links to signaling transfer points (STPs) as shown in Figure 4.8. The traffic flow between two inter-LA MTs follows the sequence: MT-BS-STP-MSC-STP-BS-MT. In this traffic pattern, the pivotal roles are played by the MSCs and STPs, which handle all control messages. The STPs are provided with required information stored in signaling control points (SCPs), which are connected to STPs via A-links.

In the current scheme of location management, two-level data hierarchy is maintained, one in the HLR and one in the VLR, to keep track of an MU who moves from cell to cell and may even cross one LA in cases of high mobility. Location management usually involves two kinds of activities, one on the part of the MT and the other on the part of the system providing the service. An MT may report as soon as it crosses an LA boundary. This reporting by the MT is called LU. Understandably, the higher the rate of cell crossing is, the greater the number of handoffs and LUs is. Consequently, the updating cost in location registration databases increases. In one way, reducing the number of handoffs (in case of uniform traffic distribution and identical cells) can be done by simply increasing the cell size. In the worst case, we can make the LA equal to the SA. Enlargement of cell size, however, is limited by the propagation characteristics and the number of available channels. Moreover, there is another constraint in terms of the call handling capacity of an MSC. In order to route an incoming call to an MT, its location with respect to the PCSN needs to be determined within a fixed time (delay constraint). Therefore, the system can initiate the search for the called MT, termed as paging, by simultaneously polling all the cells in case the allowable delay is one polling cycle.

As the size of the LA increases, the cost of paging will also increase as more cells are to be paged to find a called MT. On the other hand, reducing the size of an LA will increase the number of crossings per unit time. Hence, the cost of location update or registration will rise. Both paging and location updates consume scarce resources like wireless network bandwidth and power of MTs. Each has a significant cost associated with it. So, LA planning is to be based on a criterion that guarantees the total signaling load, which comprises paging and registration, is kept under tolerable limits. Therefore, it is characterized by the trade-off between the number of location updates and the amount of paging signaling that the PCSN has to deal with.

4.8.1.2 Methodology Used in the Two-Step Approach

As many variables and complicated constraints are involved in LA design, the method is decomposed into subproblems, namely (1) determination of optimum number of cells per LA, and (2) identification of the cells to be connected to an MSC.

Mobility Model of MTs

Since PCSN is becoming increasingly popular, there is a need for mathematical models to help understand the dynamics and analyze the performance of a PCSN. To simulate the movement of MTs there are several generic models (discussed in Chapter 2) such as the fluid flow model, diffusion model, gravity model, and random walk model. This design is independent of the mobility model used for the purpose of analysis. This approach, for the sake of simplicity, considers the fluid flow model to simulate the mobility of the roving MTs to capture the macroscopic movement behavior.

The fluid flow model assumes that the MTs are uniformly distributed on the surface of the cell, MTs arrive and depart an LA as a continuous flow of fluid, and direction of their movement is uniformly distributed over $[0, 2\pi]$. If an MT enters a cell at time zero, the initial location of the MT at time zero is assumed to be a uniformly distributed random vector on the cell's boundary. The average speed with which an MT is moving is v, the direction of the movement being measured with respect to the line leading to the center of the cell from the initial location of MT. It is tacitly assumed that the speed and the direction with which the MT is moving do not change until the MT departs the cell. In its simplest form, the model formulates the amount of traffic flowing out of a region to be proportional to the population density (ρ) within the region having a perimeter L. With these approximations, the number of boundary crossings is given by:

$$\Omega = (v\rho L / \pi)$$

The assumptions of traffic modeling and the system architecture are given below:

- Traffic generation is uniformly distributed spatially.
- Call arrivals follow a Poisson process.
- Call duration is exponentially distributed.
- Handoff and new calls are served from the same pool of available channels.
- Mobility model follows fluid flow.
- New call arrivals and call terminations are independent of the handoff traffic.

Assume that cells are identical in size; each having a radius R, and the BS is located at the center of each cell. Each BS has an omnidirectional antenna. As there are finite numbers of channels in the trunking pool of a cell, and the calls originate in a memoryless manner, the Erlang-B formula gives the probability that a call is being blocked [57]. In other words, given the radius of a cell, density of MT, average traffic per user, and the call blocking probability, the number of channels should be such that the Erlang-B formula is satisfied. Also assume free space propagation, with the path-loss exponent equal to four, and a fixed channel assignment strategy [58].

Optimization of Hybrid Cost

Using the optimum number of cells per LA from subproblem 1, the configuration of each LA, its area, and its perimeter by assigning cells to each MSC are found in an optimum manner such that the hybrid cost, comprising the inter-LA handoff cost and cable cost, is minimized. To perform the tasks in subproblem 2, a heuristic solution, in the form of a Greedy Heuristic Algorithm

(GHA) is proposed (detail of the algorithm is given in Section 4.8.1.3). From the solution of subproblem 2, the actual configuration of an LA is obtained. The two techniques used here to solve the subproblems 1 and 2 are complementary to each other, and, thus, the integrated technique optimizes both costs.

4.8.1.3   Solution Techniques Used in the Two-Step Approach

Since the designed PCSN will be overlaid on an existing PSTN serving as the backbone, MSCs will be collocated with the existing switches for PSTN wherever MSCs are required. An MT can be in either of the two states. It may be switched on or off. In the latter state, it is unreachable. This means not only that the MT does not want to make or receive any calls, but also that the network itself cannot detect the current position of the MT. An MT, which has been switched off, may move into a new LA or even into another network operator's area. When switched on again, it should inform the network about its status and location through attachment or registration.

The penetration factor p denotes the fraction of total MTs attached to a network. So, the total number of attached MTs in an SA will be Np, where N is the number of MTs in the service area. Hence, the density of attached MTs is given by:

$$\rho = Np / (S_j A_{cell})$$

where S_j is the total number of cells in the jth LA and A_{cell} the area of a cell.

The average number of handoffs h_{av} that a call will undergo during its lifetime [58] is:

$$h_{av} = [(3 + 2\sqrt{3})v] / (9\mu_{av} R_j)$$

The average call termination rate based on [57] is:

$$\mu_{total} = \mu_{av} (1 + 9h_{av})$$

Conventionally, a combination of speed and call arrival rate is considered as a single parameter, CMR [27, 59]. CMR is used to indicate the relative frequency of movement of an MT and the call arrival rate to it. It is defined as $\varphi = \mu = (\mu \sqrt{A_{cell}})/v$, which is a unitless quantity.

Paging Cost

As the MSC sends a paging message to all of the BSs under its control in order to find the called MT, each cell in an LA will carry the paging traffic associated with the called MT in all cells within that LA. Thus, the cost of paging C_p in any given cell of an LA is the product of number of cells in the LA, the incoming traffic per MT, the density of MTs, the area of a cell, and the time bandwidth product for paging.

$$C_p(Sj) = SjA_{\text{cell}} T_p B_p \mu \rho \tag{4.5}$$

where T_p is the time taken to transmit paging messages and B_p is the paging bandwidth.

LU Cost

Only a subset of S_j cells contained in the j-th LA lies on the boundary of the LA. The MTs moving out of the LA from these cells contribute to LUs in the database of HLR. So, only a shell-like area at the periphery of the LA, denoted by $L_f(S_j)$, needs to be taken into account to determine the registration cost. Using the fluid flow mobility model [60], the rate of location update is determined as $L_f(S_j) v \pi / \rho$. Registration cost in a boundary cell is then given by

$$C_r(S_j) = T_u B_u\, L_f(S_j) v\pi / \rho \tag{4.6}$$

From (4.5) and (4.6), the recurring cost, $C_t(S_j)$ in an LA comprising S_j cells becomes

$$C_t(S_j) = S_j C_p(S_j) + S_B(S_j)\, C_r(S_j)$$

where $S_B(S_j)$ is the number of cells lying on the boundary of the LA which are crossed by the roving MTs resulting in LU.

Total Recurring System Cost

The total recurring system cost is calculated due to paging and location update for the entire SA and minimized to get the optimum number of cells per LA. In the SA, there are S_T/S_j LAs. The overall recurring system cost in the entire SA, that is, the LMCF of the network over a certain time is, $C_t(S_j)(S_T/S_j)$. S_T being a constant for a certain coverage area, we minimize $C_t(S_j)/S_j$. This is equivalent to minimizing the total recurring system cost which is elaborated upon in the next section. We define $C_t(S_j)/S_j$ as normalized recurring system cost, $C_{nt}(S_j)$.

Subproblem 1

The subproblem 1 can be stated as follows:

> Given a group of cells, number of switches, average speed of MTs, density of MTs, radius of a cell, and call arrival rate, the problem is to find the optimum number of cells per LA for which the total recurring system cost is at a minimum.

Minimize $\Sigma C_t(S_j)$
or, $C_t(S_j)(S_T/S_j)$
or, $C_{nt}(S_j)$
subject to $\tau_d = 1$

Optimization of LMCF

Substituting the expressions for paging and location updates from above, we get

$$C_{nt}(S_j) = S_j A_{cell} T_p B_p \mu\rho + [S_B(S_j)/S_j] T_u B_u L_f(S_j)v\rho / \pi \qquad (4.7)$$

For large S_j, $S_B(S_j)$ is approximated as [60] $S_g(S_j) \approx c\sqrt{s_j}$

An empirical relation has been found which expresses $L_f(S_j)$ as a function of S_j [61]. That is, $L_f(S_j)/L = (b + \gamma S_j^{\kappa-1})$, where b is a constant ($b = 0.3333$), γ is a constant ($= 0.309$), and κ is a constant ($= 0.574965$). Substituting the values of $S_B(S_j)$ and $L_f(S_j)$ in (4.7), the expression for normalized recurring system cost becomes:

$$C_{nt}(S_j) = S_j A_{cell} T_p B_p \mu\rho c S_j^{-0.5} T_u B_u L(b + \gamma S_j^{\kappa-1})v\rho / \pi \qquad .$$

Stationary points in the expression of $C_{nt}(S_j)$ can be found by equating the first differential coefficient of $C_{nt}(S_j)$, with respect to S_j, to zero (i.e., $d\, C_{nt}(S_j)/\, dS_j = 0$). The stationary points become minimum for $d^2 C_{nt}(S_j)/\, dS_j^2 < 0$.
Finally, an equation like this is obtained:

$$S_j^{1.925035} - 0.1837763(T_u B_u / T_p B_p)(Lv / \mu A_{cell})S_j^{0.425}$$
$$- 0.3151671(Lv / \mu A_{cell})(T_u B_u / T_p B_p) = 0$$

S_j^{opt} can be found by solving the above equation using any standard method of numerical analysis [62]. For different call arrival rates at different times and different categories of MTs, we have minimized the total recurring system cost and thereby determined the optimum number of cells per LA. The recurring cost per call arrival $C_{pc}(S_j)$ becomes $C_{pc}(S_j) = \rho\, S_j A_{cell}\, T_p\, B_p + (S_B \sqrt{A_{cell}}\, C_r(S_j))/(\phi v\, S_j)$. For different CMRs, we can also compute $C_{nt}^{opt}(S_j^{opt})/\mu$, that is, optimum normalized recurring system cost per call arrival at different times of a day using the data of [63].

Subproblem 2

The subproblem 2 considered in this work can be stated as follows:
Given a group of cells, a group of switches whose locations are known, and an estimate of LA size, the problem is to assign cells to switches in an optimum manner such as to minimize the hybrid cost comprising the handoff cost and cost of cabling. The optimization is to be carried out subject to the following

constraints: (1) total traffic generated from the attached MTs in an SA is less than the total traffic handling capacities of all the MSCs, and (2) the traffic generated by the attached MTs in an LA is less than equal to the traffic handling capacity of the respective MSC.

The following notations are used in order to formulate the problem mathematically:

x_{ik} = an assignment variable

= 1, if cell i$(1 \leq i \leq S)$ is connected switch k, $(1 \leq k \leq Q)$

= 0, otherwise

Handoff cost, h_{ij}, is directly proportional to the frequency of handoff that occurs between cells i and j, which are assumed to be known. The sum of handoff cost of the i-th cell with all its neighbors divided by the number of neighbors gives the average handoff cost h_{av} of the i-th cell, that is, $h_{av} = \Sigma h_{ij}$/number of its neighbors. Since each cell is assigned to only one switch, we have the following constraint:

$$\Sigma_k x_{ik} = 1 \text{ for } 1 \leq i \leq S \text{ and } 1 \leq k \leq Q \tag{4.8}$$

The optimization is to be carried out in such a way that the call handling capacity of a switch is not violated. The constraint on the call handling capacity is as follows:

$$\Sigma_i \lambda_i. x_{ik} \leq 5M_k \tag{4.9}$$

A new variable y_{ij} is defined below:

y_{ji} = 1, if cell i and cell j are connected via a common switch

= 0, if cell i and cell j are connected via different switches.

Hence, handoff will occur for zero values of y_{ij}. Since the total cost = (cable cost + handoff cost), we will minimize the objective function given by:

$$\Sigma\Sigma C_{ik}.x_{ik} + \Sigma\Sigma h_{ij}(1 - y_{ij}) \tag{4.10}$$

A mathematical programming problem is defined by the objective function (4.10) subject to the constraints (4.8) and (4.9). Conventionally, the above optimization problem has been attempted by either the integer programming method [64, 65] or some heuristics [62, 64, 66]. With S cells and Q switches, the possible number of cell-to-switch assignments is S^Q, that is, the solution space is S^Q. Evidently, an exhaustive method would result in an exponential time complexity. In fact, this subproblem can be mapped to a bin-packing problem, which is NP-complete [67]. Considering the difficult optimization problem at hand, techniques that produce suboptimal solutions within acceptable run times, without searching the whole solution space, are preferred. One such method is proposed here, the GHA, based on the concept of natural clustering [68]. This is an extension of our earlier work [66] on cell-to-switch assignment. The difference lies in intelligently utilizing the estimated knowledge of the size of the LA, obtained from subproblem 1, to expedite the assignment further.

Since the number of LAs must be equal to the number of switches (because one switch is assigned per cluster), LAs will grow around the cells, which house the switches, called the seed cells [11]. The clustering of cells into an LA around a switch assumes that the cluster will grow uniformly with the MSC at its center until the cardinality of the set attains the estimated value for LA size, namely S_j^{opt}, which is obtained from the solution of subproblem 1. A seed cell is the first element of each set, assigned to an MSC. In the iteration, the number of neighboring cells of each set is determined. If the cardinality of a neighboring set is less than the estimated value S_j^{opt}, all the cells of the set are assigned to that MSC in the iteration. The switching capacity of that MSC is updated accordingly. Otherwise, the elements of a neighboring set are sorted in the descending order of handoff cost. Cells to be assigned to an MSC are chosen starting from the top of this sorted list unless (1) the number of cells assigned to the MSC reaches the estimated value, or (2) the switching capacity of the MSC is exhausted. After the iteration, the traffic handling capacity of an MSC, utilized thus far, is determined for deciding the assignment to be made during the course of next iteration. The cells common to more than one neighboring set are assigned to that set with which these cells show the maximum handoff cost. In case this handoff cost becomes equal for more than one set, cable cost is taken into account before assigning a cell to an MSC. If there is a tie, it is broken randomly. The assignment continues until the set of unassigned cells is null or the call-handling capacity of each MSC is exhausted. The pseudocode of the algorithm GHA is given below:

Description of GHA

```
Input: a) d_j , ∀j ∈ [1, Q]
    b) S_j^opt , ∀j ∈ [1, Q]
    c) M_j , ∀j ∈ [1, Q]
Output: Set of cells assigned to each switch, i.e., set_j,
    ∀j ∈ [1, Q]
Procedure:
Let m^0_j = S_j^opt, ∀j ∈ [1, Q]
FOR all j∈[1,Q] repeat the following steps until all cells
have been assigned
STEP 0: INITIAL ASSIGNMENT /* for seed cell */
```

- Initialize set_j to d_j i.e., $set^0_j = \{d_j\}$;

- Update $m_j : m^0_j = m^0_j - 1$;

- Initialize call handling capacity, M_j: $M^0_j = M^0_j - \lambda_i$;

- If $(M^i_j < 0)$ then go to ENDSTEP.

STEP i(>0): ITERATION

i.1) Identify the neighboring cells of set^{i-1}_j and find the cardinality of the neighboring set (i.e., the number of neighbors) denoted by n^{i-1}_j;

i.2) If $(m^{i-1}_j \geq n^{i-1}_j)$ then go to step i.3) else go to step i.4);

i.3) If (there is no common element between neighboring sets) then go to step i.3.1) else go to i.3.5)

i.3.1) assign all the entries to set $^{i-1}_j$;

i.3.2) update M_j : $M^i_j = M^{i-1}_j - \sum_{k=1}^{nj^{i-1}} \lambda_k$;

i.3.3) update m_j : $m^i_j = m^{i-1}_j - n^{i-1}_j$;

i.3.4) If $(M^i_j < 0)$ then go to ENDSTEP, else go to step i.1);

i.3.5) assign all the entries except common cells to set i-1$_j$;

i.3.6) assign the common elements to that set with which it gives maximum handoff; in case of a tie, calculate cable cost and assign the cells to that set with which the cells give minimum cable cost;

i.3.7) update M_j : $M^i_j = M^{i-1}_j - \sum_{k=1}^{nj^{i-1}} \lambda_k$;

i.3.8) update m_j : $m^i_j = m^{i-1}_j - n^{i-1}_j$;

i.3.9) If $(M^i_j < 0)$ then go to ENDSTEP, else go to step i.1);

i.4) If (there is no common element between neighboring sets) then go to i.4.1) else go to i.4.5)

i.4.1) arrange the neighboring set in the descending order of handoff cost ;

i.4.2) assign all m_j the entries of the sorted array to set^{i-1}_j;

i.4.3) update M_j : $M^i_j = M^{i-1}_j - \Sigma_{k=1}^x \lambda_k$, where X = n^{i-1}_j;

i.4.4) If $(M^i_j < 0)$ then go to ENDSTEP, else go to step i.1);

i.4.5) assign m_j entries except common cells to set^{i-1}_j;

i.4.6) assign the common elements to that set with which it gives maximum handoff; in case of a tie, calculate cable cost and assign the cells to that set with which the cells give minimum cable cost;

i.4.7) update M_j : $M^i_j = M^{i-1}_j - \Sigma_{k=1}^x \lambda_k$;

i.4.8) If $(M^i_j < 0)$ then go to ENDSTEP, else go to step i.1);

ENDFOR

ENDSTEP:

 If (no cell remains unassigned and $M_j \geq 0$)

 then report   SUCCESS

 else report   FAILURE

In order to design a feasible PCSN, constraints like traffic-handling capacity of MSCs and costs related to paging, registration, and cabling should be considered. Utilizing the available information of MTs and the network in a suitable manner, it is possible to devise a technique for planning LAs in a PCSN that optimizes both recurring system cost and hybrid cost. With the network engineering and configuration in mind, the fluid flow mobility model is capable of investigating the performance and cost of signaling traffic.

As this two-step approach toward LA planning takes into account both cost and network planning factors, this unique combination is of great interest to PCSN designers. It develops an optimum network planning method for a wide range of CMRs that minimizes the total recurring system cost while still ensuring good system performance. Approximate optimal results, with respect to cell-to-switch assignment, are achievable with a reasonable computational effort. However, an ideal LU and paging mechanism for high values of CMR is warranted to adjust LA per MT basis.

This design is not restricted to any particular assumption on the mobility pattern of MTs or the mobility model either. Since, the optimum LA size decreases significantly with the increase in CMR, as the corresponding hike in system cost is very high, design parameters at BSs and MSCs cannot be specified until cell allocation is completed. Finally, channel assignment, which can further improve system performance in terms of QoS and improved carrier interference ratio, can only be determined once the architecture of the PCSN has been obtained.

4.8.2 LA Planning and Signaling Requirements

In second-generation mobile systems [7], LA planning does not generate significant problems since the number of generated LUs (as well as the amount of paging signaling) remains relatively low because of the low number of users. For 3G mobile telecommunication systems, several alternative location-tracking techniques have been used (e.g., based on the use of reporting centers or a dynamic LA management protocol). Nevertheless, in UMTS an approach similar to the 2G location-finding approach has been used. That is, the system area is divided into LAs and a called user is located in two steps: (1) determine the LA within which the user roams, and (2) perform paging within this LA. The main issue concerning LA planning in 3G mobile telecommunication systems is the amount of location-finding-related signaling load [paging signaling, location updating, and distributed database (DDB) queries].

Since the size and shape of an LA affects the signaling requirements due to paging and LU, it is obvious that LA planning should minimize both, if possible. In order to provide a clear view of the relation between the LA planning and the above-mentioned parameters, we consider two extreme LA planning approaches:

1. *The system area is equal to an LA.* Whenever an MT is called, it is paged over the whole system coverage area, while no LUs are performed due to MU movements. In this case, paging signaling load can be enormous especially during the busy hour.

2. *The cell area is equal to an LA.* In this case, the location of an MT is determined with an accuracy of a single cell area. The need for paging here is minimal; paging does not locate the MT, it just alerts the terminal for the incoming call. The number of Lus, however, is expected to be enormous due to the small cell size and the user mobility. A brief description of the LA planning methods under consideration follows:

    a. *LA planning based on heuristic algorithms:* This is a method to approximate the optimum LA planning as a set of cells. According to the example heuristic algorithm used in this chapter, cells are randomly selected to form LAs.

    b. *LA planning based on area zones and highway topology:* Area zones are defined according to geographical criteria (e.g., city center), and the approach considers the population distribution and the way that people move via the city highways in order to determine the proper LA configuration.

    c. *LA planning based on overlapping LA borders:* This method can be considered an attempt to improve the previous one by means of reducing the number of generated LUs. In this case, LAs have overlapping borders in order to avoid LUs due to MU movements around the LA borders.

    d. *Time-dependent LA configuration:* According to this scenario, the network alters the LA configuration based on either some predefined timetable or monitoring of the number of LUs and the number of paging messages. The LA configuration here is selected so as to fit the time-variable mobility and traffic conditions.

    e. *LA planning based on MU grouping:* This method considers the mobility behavior of each individual MU in order to minimize the number of LUs generated due to daily MU movements. To apply this method, MUs are grouped based on their mobility behavior (e.g., high-mobility MUs) and a different LA configuration is determined for each group.

    f. *LA planning using simulated annealing* [69]: This research, focusing on LA management in wireless cellular networks, has minimized the total paging and registration cost. This chapter finds an optimal method for determining the LAs. To that end, an appropriate objective function is defined with the addition of paging and registration costs. For that purpose, the available network information to formulate a realistic optimization problem is used. In reality, the load (i.e., cost) of paging and registration to the network varies from cell to cell. An algorithm based on simulated annealing for the solution of the resulting problem is used.

4.9 Conclusion

Location management is a key factor for wireless mobile networks. Without a good strategy for location management, mobile communication and computing cannot exist. Mobile cellular radio networks of the third generation will be used by an increasing number of users. Thus, location management functions for providing a continuity of service over the whole system area have to be much more sophisticated than in today's cellular networks.

In this chapter, we have looked at different mechanisms to locate an MT's current network address and discussed their advantages and disadvantages. Location management functions, such as location updating and paging, have to fulfill services and the requirements of users and operators. One of these requirements is cost efficiency, which could be reached by minimizing the signaling traffic both on radio links and on fixed network links. What we aim for is a location management scheme that will provide efficient searches and updates transparent to the user.

References

[1] Madhow, U., M. L. Honig, and K. Steiglitz, "Optimization of Wireless Resources for Personal Communications Mobility Tracking," *IEEE/ACM Trans. on Networking*, Vol. 3, No. 6, 1995, pp. 698–706.

[2] Ramanathan S., and Martha Steenstrup, "A Survey of Routing Techniques for Mobile Communications Networks," 1996.

[3] Ketchum, J. W., "Routing in Cellular Mobile Radio Communication Networks," *Routing in Communication Networks*, M. Steenstrup (ed.), Englewood Cliffs, NJ: Prentice-Hall, 1995.

[4] Mouly, M., and M. B. Pautet, "The GSM System for Mobile Communications," Palaiseu, France: M. Mouly, 1992.

[5] Telecommunications Industry Association, Cellular Radio Telecommunication Intersystem Operation, TIA/EIA IS-41B, 1991.

[6]   Mohan, S., and R. Jain, "Two User Location Strategies for Personal Communications Services," *IEEE Personal Communications*, First Quarter, 1994, pp. 42–50.

[7]   G. L., Markoulidakis, et al., "Evaluation in LA Planning in Future Mobile Telecommunication Systems," *Wireless Network*, 1995.

[8]   Bar-Noy, A., and I. Kessler, "Tracking Mobile Users in Wireless Communications Networks," *IEEE Trans. on Information Theory*, Vol. 39, No. 6, 1993, pp. 1877–1886.

[9]   Xie, H., S. Tabbane, and D. J. Goodman, "Dynamic Location Area Management and Performance Analysis," *Proc. 43rd IEEE Vehicular Tech. Conf.*, 1993, pp. 536–539.

[10]   Birk Y., and Y. Nachman, "Using Direction and Elapsed-Time Information to Reduce the Wireless Cost of Locating Mobile Units in Cellular Networks," *Wireless Networks*, Vol. 1, No. 4, 1995, pp. 403–412.

[11]   Bar-Noy, A., I. Kessler, and M. Sidi, "Mobile Users: To Update or Not To Update?" *Wireless Networks*, Vol. 1, No. 2, 1995, pp. 175–186.

[12]   Rose, C., and R. Yates, "Minimizing the Average Cost of Paging Under Delay Constraints," *Wireless Networks*, Vol. 1, Feb. 1995, pp. 211–219.

[13]   Akyildiz, Ian F., and Joseph S. M. Ho, "A Mobile User Location Update and Paging Mechanism Under Delay Constraints, *Proc. of ACM SIGCOMM*, Cambridge, MA, 1995, pp. 244–255.

[14]   IETF Mobile-IP Working Group, IPv4 Mobility Support, working draft, 1995.

[15]   Meier-Hellstern, K., and E. Alonso, "The Use of SS7 and GSM to Support High Density Personal Communications," *Proc. of ICC*, 1992, pp. 1698–1702.

[16]   Lo, V. N., R. S. Wolf, and R. C. Bernhardt, "Expected Network Database Transaction Volume to Support Personal Communications Services," *1st International Conference Universal Personal Communications Services*, Dallas, TX, 1992.

[17]   Jain, R., et al., "A Caching Strategy To Reduce Network Impacts of PCS," *IEEE J. on Selected Areas in Communications*, Vol. 12, No. 8, 1994, pp. 1434–1444.

[18]   Ho, J. S. M., and I. F. Akyildiz, "Local Anchor Scheme for Reducing Location Tracking Costs in PCNs," *Proc. of ACM MOBICOM*, Berkeley, CA, 1995, pp. 181–194.

[19]   Awerbuch, B., and D. Peleg, "Concurrent Online Tracking of Mobile Users," *Proc. of ACM SIGCOMM*, Zurich, Switzerland, 1991, pp. 221–234.

[20]   Anantharam, V., et al, "Optimization of a Database Hierarchy for Mobility Tracking in a Personal Communications Network," *Performance Evaluation*, Vol. 20, 1994, pp. 287–300.

[21]   Badrinath, B. R., T. Imielinski, and A. Virmani, "Locating Strategies for Personal Communication Networks," *Proc. of Workshop on Networking of Personal Communications Applications*, 1992.

[22]   Ozsu, M. T., and P. Valduriez, "Principles of Distributed Systems," Englewood Cliffs, NJ: Prentice-Hall, 1991.

[23]   Dowdy, L. W., and D. V. Foster, "Comparative Models of the File Allocation Problem," *ACM Computing Surveys*, Vol. 14, No. 2, 1982, pp. 287–313.

[24] Shivakumar, N., and J. Widom, "User Profile Replication for Faster Location Lookup in Mobile Environments," *Proc. of ACM MOBICOM*, Berkeley, CA, 1995, pp. 161–169.

[25] Akyildiz, Ian F., and Joseph S. M. Ho, "On Location Management for Personal Communications Networks," *IEEE Communications Magazine, Sept.* 1996, pp. 138–45.

[26] Rose, C., "Minimizing the Average Cost of Paging and Registration: A Timer-Based Method," *ACM J. Wireless Networks*, Feb. 1996, pp. 109–16.

[27] Akyildiz, Ian F., Joseph S. M. Ho, and Yi-Bing Lin, "Movement-Based Location Update and Selective Paging for PCS Networks," 1996.

[28] Bar-Noy, A., and I. Kessler, "Mobile Users: To Update or Not To Update?" *Proc. INFOCOM 94*, June 1994, pp. 570–76.

[29] Lyberopoulos, G. L., et al., "Intelligent Paging Strategies for Third Generation Mobile Telecommunication Systems," IEEE Trans on Vehicular Technology, Aug 1995, pp. 543-53.

[30] Wang, Wenye, Ian F. Akyildiz, and Gordon L. Stüber, "Effective Paging Schemes with Delay Bounds As QoS Constraints in Wireless Systems," *Wireless Networks*, Vol. 7, 2001, pp. 455–466.

[31] Bhattacharjee, P. S., D. Saha, and A. Mukherjee, "Paging Strategies for Future Personal Communication Services Networks," *Proc 6th Int. Conf. on High Performance Computing (HiPC'99)*, Calcutta, India, Dec. 1999.

[32] Papoulis, A., *Probability, Random Variable, and Stochastic Processes, 3rd Edition*, New York: McGraw Hill, 1991.

[33] Wang, Wenye, Ian F. Akyildiz, and Gordon L. Stüber, "Reducing the Paging Costs Under Delay Bounds for PCS Networks," *Proc. of IEEE WCNC'2000*, Sept. 2000.

[34] Akyildiz, Ian F., and Wenye Wang, "A Dynamic Location Management Scheme for Next-Generation Multitier PCS Systems," *IEEE Transactions on Wireless Communications*, Jan. 2002.

[35] Sarikaya, Behcet, and Sridhar Gurivireddy, "Evaluation of CDMA2000 Support for IP Micromobility Handover and Paging Protocols," *IEEE Communications Magazine*, May 2002, pp. 146–149.

[36] Kempf, J., et al., "Requirements for Layer 2 Protocols to Support Optimized Handover for IP Mobility," *IETF Draft*, July 2001.

[37] Kempf, J., et al., "Bidirectional Edge Tunnel Handover for IPv6," *IETF draft*, Sept. 2001.

[38] Gurivireddy, S., B. Sarikaya, and A. Krywaniuk, "Layer-2 Aided Mobility Independent Dormant Host Alerting Protocol," *IETF draft*, Sept. 2001.

[39] EIA/TIA, "Cellular Radio-Telecommunications Intersystem Operations," Tech. Rep. IS-41 Revision C, 1995.

[40] Tabbane, Sami, "Location Management Methods for Third-Generation Mobile Systems," *IEEE Communications Magazine*, Aug. 1997, pp. 72–84.

[41] Lee, W. C. Y., *Mobile Cellular Telecommunications Systems*, New York: McGraw-Hill, 1989.

[42] Wang, D. C. C., "A Survey of Number Mobility Techniques for PCS," *Proc. IEEE Int. Conf. On Personal Communications*, Tokyo, Japan, Nov. 6–10, 1995.

[43] Wang, D. C. C., "A Survey of Number Mobility Techniques for PCS," *Proc. IEEE ICC*, 1994.

[44] Tabbane, S., "Database Architectures and Location Strategies for Mobility Management in Mobile Radio Systems," *Proc. Wksp. Multiaccess, Mobility and Teletraffic for Persona. Communications,* Paris, France, May 1996.

[45] Jabbari, B., "Intelligent Network Concepts in Mobile Communications," *IEEE Communication Magazine*, Feb. 1992.

[46] Plassmann, D., "Location Management for MBS," *Proc. IEEE VTC*, Stockholm, Sweden, June 8–10, 1994, pp. 649–53.

[47] Okasaka, S., et al., "A New Location Updating Method for Digital Cellular Systems," *Proc. IEEE VTC '91*, Saint Louis, MI, May 1991.

[48] Hu, L. R. and S. S. Rappaport, "An Adaptive Location Management Scheme for Global Personal Communications," *Proc. IEEE Int. Conf. On Personal Communications*, Tokyo, Japan, Nov. 6–10, 1995.

[49] RACE II Deliverable, "Location Areas, Paging Areas and the Network Architecture," R2066/PTTNL/MF1/DS/P/001/b1, Apr. 1992.

[50] Tabbane, S., "Comparison between the Alternative Location Strategy (AS) and the Classical Location Strategy (CS)," *WINLAB Tech. Rep. 37*, Aug. 1992.

[51] Tabbane, S., "An Alternative Strategy for Location Tracking," *IEEE JSAC*, Vol. 13, No. 5, June 1995.

[52] Perkins, C., "IP Mobility Support for IPv4, revised," *draft-ietf-mobileip-rfc2002-bis-08.txt, IETF*, Sept. 2001.

[53] Misra, Archan, et al., "IDMP-Based Fast Handoffs and Paging in IP-Based 4G Mobile Networks," *IEEE Communications Magazine*, March 2002, pp. 138–145.

[54] Chan, P. M. L., et al., "Mobility Management Incorporating Fuzzy Logic for a Heterogeneous IP Environment," *IEEE Communications Magazine,* Dec. 2001, pp. 42–51.

[55] Bhattacharjee, P. S., et al., "A Practical Approach for Location Area Planning in a Personal Communication Services Network," *Proceedings of MMT '98*, USA, 1998.

[56] Modarressi, A. R., and R. A. Skoog, "Signaling System No. 7: A Tutorial," *IEEE Communication Magazine*, July 1990 pp. 19–35.

[57] Gunz, A., C. M. Krishna, and D. Tang, "On Optimal Design of Multitier Wireless Cellular Systems," *IEEE Communications Magazine*, Feb. 1997, pp. 88–93.

[58] Guerin, R., "Channel Occupancy Time Distribution in a Cellular Radio System," *IEEE Trans. on Vehicular Tech.*, Vol. VT-35, No. 3, Aug. 1987.

[59] Chih-Lin, I., P. Pollini, and R. B. Gitlin, "PCS Mobility Management Using Virtual Call Setup Algorithm," *IEEE Trans. on Networking*, Vol. 5, No. 1, Feb. 1997.

[60] Alonso, E., K. S. Hellstern, and G. P. Pollini, "Influence of Cell Geometry on Handover and Registration Rates in Cellular and Universal Personal Telecomm Networks," *Proc. 8th Int. Teletraffic Seminar*, Sonata Marghareta, Ligure, Geneva, Italy.

[61] Bhattacharjee, P. S., D. Saha, and A. Mukherjee, "Heuristics for Assignment of Cells to Switches in a PCSN: A Comparative Study", *ICPWC'99, Proc. IEEE International Conference Personal Communication*, Jaipur, India, 1999, pp. 331–334.

[62] Jain, M. K., S. R. K. Iyengar, and R. K. Jain, *Numerical Methods for Scientific and Engineering Computation*, New York: Wiley Eastern Limited, 1985, pp. 41–43.

[63] Markoulidakis, J. G., G. L. Lyberopolous, and M. E. Anynostom, "Traffic Model for Third-Generation Cellular Mobile Telecomm. Systems," *Wireless Network*, 1998.

[64] Merchant, A., and B. Sengupta, "Assignment of Cells to Switches in PCS Networks," *IEEE/ACM Trans. on Networking*, Vol. 3, No. 5, Oct. 1995, pp. 521–526.

[65] Nemhauser, G. L., and L. A. Woolly, *Integer and Combinatorial Optimization*, New York: Wiley, 1988.

[66] Saha, D., A. Mukherjee, and P. S. Bhattacharjee, "A Simple Heuristic for Assignment of Cells to Switches in a PCS Network," *Wireless Personal Communication*, Netherlands: Kluwer Academic Publishers, 2000, pp. 209–224.

[67] Gary M. R., and D. S. Johnson, "Computers and Intractability, A Guide to the Theory of NP-Completeness," New York: W. H. Freeman, 1979.

[68] Saha, D., and A. Mukherjee, "Design of Hierarchical Communication Network Under Node-Link Failure Constraints," *Computer Communication*, Vol. 18, No. 5, 1995, pp. 378–383.

[69] Demirkol, Ilker, et al., "Location Area Planning in Cellular Networks Using Simulated Annealing," *Proc. of INFOCOM*, 2001.

Part II
Ad Hoc Wireless Networks

5

Overview

5.1 Characteristics of Ad Hoc Networks

Ad hoc networks have several salient characteristics [1–4]:

- *Dynamic topologies:* Because of the possibly rapid and unpredictable movement of the nodes and fast-changing propagation conditions, network information, such as link-state, becomes quickly obsolete. This leads to frequent network reconfigurations and frequent exchanges of control information over the wireless medium.

- *Asymmetric link characteristics:* In a wireless environment, communication between two nodes may not work equally well in both directions. In other words, even if node n is within the transmission range of node m, the reverse may not be true. Although we have assumed bidirectional links in Figure 1.5, some of them may be unidirectional in a real-life scenario.

- *Multihop communication:* Each node in an ad hoc network will act as a transmitter, a receiver, or a relay station. So, packets from a transmitter node (source) may reach the receiver node (destination) in multiple hops through several intermediate relay nodes. However, the successful operation of an ad hoc network will be hampered if an intermediate node, participating in a communication between a source-destination pair, moves out of range suddenly or switches itself off in between message transfer. The situation is worse if there is no other path between those two nodes. In Figure 1.5, if D moves out of range disconnecting the link between C and D, or if D switches itself off, the communication between *C* and *F* would be interrupted. The absence of *D* creates two disconnected components: {*A*, *B*, *C*} and {*E*, *F*, *G*}.

- *Decentralized operation:* Ad hoc networks are network architectures that can be rapidly deployed and that do not need to rely on preexisting infrastructure or centralized control. In cellular wireless networks, there are a number of centralized entities; (e.g., BSs, MSCs, and the HLR) In ad hoc networks, since there is no preexisting infrastructure, these centralized entities do not exist. The centralized entities in the cellular networks perform the function of coordination. Thus, lack of these entities in the ad hoc networks requires more sophisticated distributed algorithms to perform equivalent functions.

- *Bandwidth-constrained variable-capacity links:* Wireless links will continue to have significantly lower capacity than their hardwired counter parts. In addition, the realized throughput of wireless communications is often much less than a radio's maximum transmission rate because of the effects of multiple access, fading, noise, and interference conditions, for example. One effect of the relatively low to moderate link capacities is that congestion is typically the norm rather than the exception. Thus, the aggregate application demand will likely approach or exceed network capacity frequently. This demand will continue to increase as multimedia computing and collaborative networking applications increase.

- *Energy-constrained operation:* The mobile nodes in an ad hoc network rely on batteries or other exhaustible means for their energy. For these nodes, the most important system design criteria for optimization may be energy conservation. One way of achieving this is to optimize the transmission power of each node.

These characteristics of ad hoc networks create a set of performance concerns for protocol design that extend beyond those guiding the design of protocols for conventional networks with preconfigured topology.

5.2 Three Fundamental Design Choices

We focus here on three fundamental choices in the design of ad hoc networks [5]:

1. The network architecture;
2. The routing protocol;
3. The medium access control.

5.2.1 Flat Versus Hierarchical Architecture

The architecture of ad hoc networks can be classified into hierarchical and flat architecture [5]. In a hierarchical architecture, the network nodes are dynamically partitioned into groups called clusters. Thus, the details of the network topology are concealed by aggregating nodes into clusters, clusters into super-clusters, and so on [6]. The membership in each cluster changes over time in response to node mobility and is determined by the criteria specified in the clustering algorithm. Within each cluster, one node is chosen to perform the function of a cluster head [7, 8]. Routing traffic between two nodes that are in two different clusters is normally done through the cluster heads of the source and destination clusters. A dynamic cluster head selection algorithm is used to elect a node as the cluster head. However, frequent change in cluster membership and consequent reselection of cluster head adversely affect routing protocol performance. In order to get rid of this problem, a fully distributed approach for cluster formation and intracluster communication has been proposed [9] that eliminate the requirement for a cluster head altogether. Even then, in the context of a highly dynamic scenario, the reconfiguration of clusters and the assignment of nodes to clusters do require excessive processing and increase communication overhead. This is depicted in Figure 5.1, where three clusters are shown. Here, nodes marked *C* are cluster heads and nodes marked *N* are cluster members. Another type of node, called a gateway node, is introduced here. Nodes marked

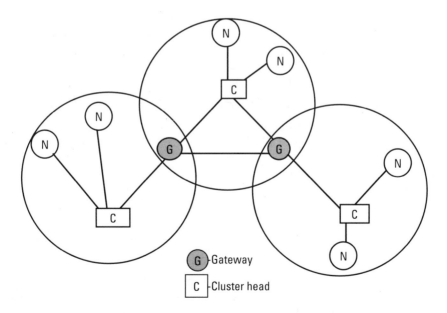

Figure 5.1 Clustered architecture of ad hoc networks.

G are gateway nodes for intercluster communications. Intracluster communications are done through the respective cluster head. The detailed routing mechanism in this type of architecture will be illustrated in Chapter 7.

In contrast, in a flat architecture, there are no clusters and the neighboring nodes can communicate directly. It has been argued that the routing in flat architecture is more optimal (close-by nodes do not have to communicate through the hierarchy) and the network tends to better balance the load among multiple paths, thus reducing the traffic bottlenecks that occur at the cluster nodes in the hierarchical approach [5].

In a hierarchical architecture, some nodes, such as cluster heads and gate way nodes, have a higher computation and communication burden than other nodes. The network reliability may be affected due to this single point of failure of these critical nodes. However, the routing messages may only have to propagate within a cluster. Thus, the number of globally propagated routing messages is small. On the contrary, in a flat architecture, all nodes carry the same responsibility, and reliability is not dependent on any single point of failure. However, this flat architecture is not bandwidth efficient because the routing messages have to propagate globally throughout the network. The scalability gets worse when the number of nodes increases.

The main advantage of the hierarchical ad hoc network is the ease of the mobility management process. In order to limit far-reaching impact to topology dynamics, complete routing information is maintained only for intracluster routing. Intercluster routing is achieved by hiding the topology details within a cluster from external nodes and using hierarchical aggregation, reactive routing, or a combination of both techniques. Cluster heads can act as a databases that contain the location of the nodes in their own clusters. To determine the existence and the location of a mobile node, a query is broadcasted to all the cluster heads. The cluster under which the node resides, responds to the query originator. A mobility management scheme needs to be implemented in the flat networks by appropriate routing algorithms, although it would increase the overhead due to control message propagation throughout the network [5].

5.2.2   Proactive Versus Reactive Routing

The existing routing protocol can be classified either as proactive or as reactive. In proactive protocols the routing information within the network is always known beforehand through continuous route updates. The distance vector and link-state protocols are examples of proactive schemes. Examples of proactive routing methods in an ad hoc network environment are [10, 11]. However, these methods require one to know the topology of the entire network and this information needs to be propagated through the network. As the network size increases and as the nodal mobility increases, a smaller and smaller fraction of this total amount of control traffic will be used. This is so, since as the nodes become more mobile, the lifetime of a link decreases. Thus, the period for which the routing information remains valid decreases as well. In fact, it is easy to show that, for any given network capacity, there exists a network size and nodal mobility for which all the network capacity will be wasted on control traffic only [5]. Thus, in a highly dynamic environment, these schemes are less efficient. However, the advantage of the proactive schemes is that, once a route is needed, it is immediately available from the route table.

Reactive protocols, on the other hand, invoke a route discovery procedure on demand only. Thus, when a route is needed, some sort of flooding-based global search procedure is employed. The family of classical flooding algorithms belongs to this group. Examples of reactive protocols in the context of ad hoc networks are [4, 12]. In these protocols, because route information may not be available at the time a route request is received, the delay to determine a route can be quite significant. Furthermore, the global search procedure of the reactive protocols generates significant control traffic.

Because of this long delay and excessive control traffic, pure reactive routing protocols may not be applicable to real-time communication. However, pure proactive schemes are likewise not appropriate for the ad hoc network environment, as they continuously use a large portion of the network capacity to keep the routing information current. Thus, designing a proper routing scheme for ad hoc networks is a challenging task. This will be discussed in Chapter 7.

5.2.3 Medium Access Protocol

Designing an efficient and effective medium access control (MAC) protocol with collision avoidance capability in a mobile ad hoc network is a very challenging task because the network is self-organizing without the aid of a wired backbone or a centralized network control. Moreover, the medium in a wireless network is, by nature, a shared resource where a sender normally uses omnidirectional broadcast mode to transmit a message for its intended destination. As a result, other users who are within the transmission range of the sender will also have to listen to its message even though they are not the intended receivers of this message. In this context, it is important to ensure a collision-free message communication environment even when multiple senders want to communicate with multiple receivers using the same shared medium. Imagine a situation (see Figure 5.2) with seven nodes (*I, E, S, D, H, J* and *N*) where a sender *E is trying to send a message to a receiver I,* and another sender *H* is trying to send a message to another receiver *J* simultaneously. Let us also assume that *I* is within the transmission range of *E* only and *J* is within the transmission range of *H* only. In other words, *I* can only listen to *E* and not to *H*. Similarly, *J* can only listen to *H and not to E.* This is a conflict-free communication

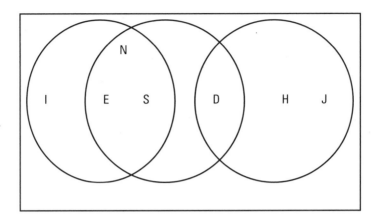

Figure 5.2 Hidden terminal and exposed terminal problems.

environment where both communications can progress simultaneously without interfering with each other. However, this scenario is rare in the dynamic environment of an ad hoc network. Let us further assume that while E is talking to I, another node S wants to send a message to N. The node N is in the listening range of both E and S. So, there will be collision at receiver N, and N will not be able to receive any message from S.

One way to solve the problem is to "sense" the medium before transmitting. So, S will first "sense" the medium to find out the existence of any ongoing communication. Since E is already transmitting data and S is within the transmission range of E, S can "sense" that the medium is busy. So, S will defer its desire to transmit a message towards N. However, this will give rise to what is known as the *exposed terminal problem*. Assume that S is transmitting to D. Since E senses an ongoing data transmission (i.e., E is exposed to the transmission by S), E remains silent. But E does not know that D is out of its reach. In fact, E could have transmitted to node I because these transmissions would not cause any collision either at D or at I.

A more serious problem is known as the *hidden terminal problem*. Assume that node S is sending data to node D. A terminal H is hidden when it is far away from the data source S but is close to the destination D. Without the ability to detect the ongoing data transmission, H will cause a collision at D if H starts transmitting to J. MAC, in this context, implements mechanisms to prevent the collision at D by any hidden terminal (such as H).

Most of the MAC protocols that tend to solve the hidden terminal and the exposed terminal problems do not address the issue of node mobility. For example, let us assume that S is transmitting to E and node N is initially outside the range of S and E. So, node N will not hear the communication and start transmitting to another node away from S and E. If N is static, this will not cause any interference. But let us assume that *N* is moving and coming closer to *S* and *E*, as shown in Figure 5.2. If *N* as a transmitter moves into the communication range of *E*, any transmission from *N* will collide with the ongoing communication between *S* and *E*, and thus affect the channel throughput. This is called an intruding terminal problem.

The main purpose of MAC mechanisms in the context of ad hoc networks is to ensure collision-free communication among nodes under different conditions. These schemes, however, waste a large portion of the network capacity by reserving the wireless media over a large area. In other words, when *S* and *D* are communicating, a large number of nodes in the neighborhood of *S* and *D* have to sit idle, waiting for the data communication between *S* and *D* to finish. To improve medium utilization, the use of directional antennas that can largely reduce the radio interference, thereby improving the utilization of wireless medium and consequently the network throughput, has been proposed. For example, when *S* and *D* are communicating with directional antennas at *S* and *D* directed towards each other, node *E* is free to communicate with anyone in its neighborhood except *S*. Similarly, node *H* can communicate in all other direction except towards *D*. However, in the context of ad hoc networks, it is difficult to find ways to control the direction of adaptive antennas for transmission and reception in each terminal in order to achieve an effective multihop communication between any source and destination. This difficulty is mainly due to mobility and lack of centralized control in ad hoc networks. Thus, developing a suitable MAC protocol in ad hoc networks using adaptive antennas is a challenging task.

Another way to improve medium utilization is to use mechanisms of controlling the transmission power of senders. For example, a sender may reduce its current transmission range to a minimum value that is sufficient for the successful reception of its message at the intended destination. Medium utilization can be significantly increased because the severity of signal overlapping is reduced. However, adaptive power control mechanisms introduce additional overhead and problems of route discovery and topology maintenance in ad hoc networks. We will investigate these issues in the next chapter.

References

[1]   Corson, S., J. Macker, and S. Batsell, "Architectural Considerations for Mobile Mesh Networking," *Internet Draft RFC Version 2*, May 1996.

[2]   Corson, M. S., and J. Macker, "Mobile Ad Hoc Networking (MANET): Routing Protocol Performance Issues and Evaluation Considerations," *Request For Comments 2501*, Internet Engineering Task Force, Jan. 1999.

[3]   Johnson, D., "Routing in Ad Hoc Networks of Mobile Hosts," *Proc. IEEE Workshop on Mobile Comp. Systems and Appls.*, Dec. 1994.

[4]   Johnson, B., and D. A. Maltz, "Dynamic Source Routing in Ad Hoc Wireless Networks," T. Imielinski and H. Korth (eds.), *Mobile Computing*, Netherlands: Kluwer, 1996.

[5]   Haas, Z. J., and S. Tabrizi, "On Some Challenges and Design Choices in Ad Hoc Communications," *IEEE MILCOM'98*, Bedford, MA, Oct. 18–21, 1998.

[6]   Lauer, G. S., "Packet-Radio Routing," M. E. Steenstrup (ed.), *Routing in Communications Networks*, Englewood Cliffs, NJ: Prentice Hall, 1995, pp. 375–379.

[7]   Baker, D. J., and A. Ephremides, "The Organization of a Mobile Radio Network Via a Distributed Algorithm," *IEEE Trans. Commun.*, Vol. COM-29, 1981.

[8]   Gerla, M., and J. T. Tsai, "Multicluster, Mobile, Multimedia Radio Network," *ACM Wireless Networks*, Vol. 1, No. 3, 1995, pp. 255–265.

[9]   Lin, R., and Mario Gerla, "Adaptive Clustering for Mobile Wireless Networks," *IEEE Journal on Selected Areas in Communications*, Vol. 15, No. 7, Sept. 1997.

[10]   Krishna, P., et al., "A Cluster-Based Approach for Routing in Ad Hoc Networks," *Proc. of the 2nd USENIX Symp. on Mobile and Location-Independent Computing*, Apr. 1995.

[11]   Perkins, E., and P. Bhagwat, "Highly Dynamic Destination-Sequenced Distance-Vector Routing (DSDV) for Mobile Computers," *ACM Comput. Commun. Rev.*, Vol. 24, No. 4, (ACM SIGCOMM'94) Oct. 1994, pp. 234–244.

[12]   Perkins, C. E., and E. M. Royer, "Ad Hoc On-Demand Distance Vector Routing," *Proc. of the 2nd IEEE Workshop on Mobile Comp. Sys. And Apps.*, Feb. 1999, pp. 90–100.

6

MAC Techniques in Ad Hoc Networks

6.1 MAC Protocols with Omnidirectional Antennas

When nodes in an ad hoc network use omnidirectional antennas, it is implied that the transmitted signal from any node will spread equally well in all directions within its transmission range. Similarly, as a receiver, a node will receive signals equally well from all directions. Traditionally, ad hoc networks use omnidirectional antennas for their simplicity and lower cost. However, the use of directional antennas can largely reduce the radio interference, thereby improving the utilization of wireless medium and consequently the network throughput. So, researchers are also exploring the possibilities of using low-cost directional antennas in the context of ad hoc networks. In this section, MAC protocols with omnidirectional antennas will be illustrated. In the next section, MAC protocols with directional antennas will be discussed.

   Designing an efficient and effective MAC protocol with collision avoidance capability in a mobile ad hoc network is a very challenging task because the network is self-organizing without the aid of wired backbone or centralized network control. Some of the existing MAC protocols can be classified into ALOHA [1], Carrier Sense Medium Access (CSMA) [2], Busy Tone Multiple Access (BTMA) [3], Multiple Access with Collision Avoidance (MACA) [4], Media Access Protocol for Wireless LANs (MACAW) [5], Floor Acquisition Multiple Access (FAMA) [6], and Dual Busy Tone Multiple Access (DBTMA) [7]. In ALOHA [1], each user can start a transmission at any time without first sensing the channel status. The throughput is affected by frequent collisions. The slotted ALOHA protocol achieves a channel utilization of 36% by dividing the channel into time slots. Transmission is allowed to start at the beginning of a time slot only. Slotted ALOHA maintains a higher channel utilization by avoiding collisions in the middle of the data transmission, but it introduces somewhat of a longer access delay and more complexity due to the need for slot synchronization. CSMA protocols have been used in a number of packet-radio networks in the past [2]. In CSMA, each user senses the carrier before starting the transmission. The throughput is higher than ALOHA because collisions may be avoided. However, the CSMA protocols do not address the hidden terminal problem and the exposed terminal problem, as illustrated in Section 5.2.3.

The BTMA [3] was the first proposal to combat the hidden terminal problems of CSMA. However, BTMA is designed for station-based networks and requires dividing the channel into a message channel and the busy tone channel. One of the first protocols for wireless networks based on a handshake between sender and receiver was the Split-Channel Reservation Multiple Access (SRMA) [8]. It uses the Request-to-Send/Clear-to-Send (RTS/CTS) dialog as a mechanism of handshaking. RTS/CTS dialog precedes the actual transmission of the data by the stations and allows the reserving of the channel, so that collisions with other stations are avoided. The MACA [4] and MACAW [5] protocols use similar schemes in single channel networks. The protocols based on an RTS/CTS-based handshaking mechanism can reduce the packet loss due to hidden terminals. An explanation follows.

Let us refer to Figure 5.2. Before sending the data packets, S sends an RTS with proposed duration of data transmission to inform its neighbors about its willingness to start a communication with D. So, all the neighbors of S will become idle (neither transmit nor receive) during this period. If the intended destination D hears the RTS, D replies with CTS to inform its neighbors about its willingness to receive data from S. The neighbors of D that receives only CTS cannot transmit (to avoid collision with D's reception), but they can receive from other nodes outside the RTS/CTS boundaries. In some protocols, S issues another packet called a data-sending packet after it receives the CTS packet in order to ensure its neighbors that a successful reservation has been accomplished. This avoids unwanted waiting of nodes receiving only RTS under the condition of unsuccessful negotiation between S and D. After data reception by D is over successfully, it issues an acknowledgement (ACK) to S. The ACK packet at link level speeds up packet retransmission that is faster than relying on the slow recovery at the transport layer.

Any other nodes overhearing the RTS must be close to node S and therefore should remain silent for a time period long enough so that node S can receive the returning CTS without any collision. Any other nodes overhearing the CTS must be close to node D and therefore should refrain from transmission for a time period that is long enough for the transmission of the proposed data packet so that node D can receive the returning data packet without any conflict. A hidden terminal (e.g., node H), which is in the range of node D but out of the range of node S, will hear the CTS but not the RTS. It therefore remains silent during the data transmission from node S to node D.

In the IEEE's proposed standard for WLANs (IEEE 802.11) [9], it describes specifications on the parameters of both the physical (PHY) and MAC layers of the network. The PHY layer, which actually handles the transmission of data between nodes, can use either direct sequence spread spectrum, frequency hopping spread spectrum, or infrared (IR) pulse position modulation. The initial specification of IEEE 802.11 makes provisions for data rates of either 1 Mbps or 2 Mbps, and calls for operation in the 2.4- to 2.4835-GHz frequency band, which is an unlicensed band for industrial, scientific, and medical (ISM) applications. The current version of IEEE 802.11b operates at 11 Mbps in the 2.4 GHz band and communicates up to 150 feet. IEEE 802.11a operates at 40 Mbps in the 5.8 GHz band. For IR transmission, it operates at 300 to 428,000 GHz. Infrared transmissions require absolute LOS links (no transmission is possible outside any simply connected space or around corners), as opposed to RF transmissions, which can penetrate walls.

The basic medium access method for IEEE 802.11 relies on the Distributed Coordination Function (DCF), which uses CSMA/Collision Avoidance with RTS/CTS-data-ACK, as discussed before. A node willing to transmit senses the medium first to determine if it is idle. If it is busy, the node waits until transmission stops and then enters into a random back-off procedure by setting an internal counter. This back-off counter determines the amount of time the node must wait until it is allowed to transmit its packet. During periods when the channel is clear, the node willing to transmit decrements its back-off counter. When the back-off counter reaches zero, the node starts transmitting packets if the medium is still idle. Since the probability that two nodes will choose the same back-off factor is small, this prevents multiple nodes from seizing the medium immediately after completion of the preceding transmission, and collisions between packets are minimized.

It is to be noted that CSMA/Collision Detection, which is commonly used in wired LANs, is impractical in this context, because WLAN radios are half duplex and cannot receive while transmitting. Therefore, a collision cannot be detected by a radio while transmission is in progress.

The basic RTS/CTS mechanism described above is supposed to prevent all other nodes in the receiver's range from transmitting. However, CTS packets can still be destroyed by collisions. It has been shown [7] that the probability of CTS packet collision in a multihop network that uses the basic RTS/CTS dialogue rules can be as high as 60% when the network operates at high traffic load. To overcome this, the DBTMA scheme [7] is proposed. It is based on both the RTS/CTS dialogue and the carrier-sensing feature. In particular, the carrier sensing is performed by the introduction of two busy tones, which indicate the status of the shared channel in a particular geographical area. This significantly reduces the chances of destruction of the actual data packets due to transmission collisions, which further improves the scheme's utilization. In this protocol, the single common channel is split into two subchannels: a data channel and a control channel. Data packets are transmitted on the data channel, while control packets (RTS and CTS, for example) are transmitted on the control channel. Additionally, two busy tones are assigned to the control channel: the receive busy tone, which shows that a node is receiving on the data channel; and the transmit busy tone, which shows that a node is transmitting on the data channel. The DBTMA scheme significantly improves the performance over the basic RTS/CTS-based schemes. However, the use of a separate channel just to convey the state of the data channel may not be a desirable feature in the context of an ad hoc wireless environment.

FAMA [6] is another proposal to improve basic RTS/CTS-based protocol. FAMA permits a sender to acquire control of the channel in the vicinity of a receiver dynamically, before transmitting data packets. The floor acquisition strategy uses an RTS/CTS handshake and is based on a few simple principles: (1) making the senders listen to the channel before transmitting RTSs; (2) implementing a busy tone mechanism using a single channel and half-duplex radios by making the receiver send CTSs that last long enough for the hidden senders to realize that they must back off; and (3) providing priority to those stations who successfully complete a handshake. FAMA is the first single-channel protocol to provide the equivalent functionality of a busy tone solution, and it substantially improves the performance by avoiding collisions.

MACA, MACAW, FAMA, and similar protocols depend on RTS/CTS dialogue to solve the hidden terminal and the exposed terminal problems. However, they do not address the issue of node mobility. The fact that a mobile node did not hear a RTS/CTS dialog does not indicate that the channel is indeed free for use. For example, if a node not hearing the RTS/CTS dialogue moves into the communication range of a receiver, any transmission of the intruder will collide with the ongoing one, and thus affect the channel throughput. This is called an intruding terminal problem [10]. In order to solve the intruding terminal problem, different spread spectrum channels for control messages (i.e., RTS and CTS) and data messages to different nodes have been proposed in [10]. Thus, an ongoing data transmission is safeguarded against collisions due to the control or data message from an intruder.

6.2   MAC Protocols with Directional Antennas

The RTS/CTS-based scheme using omnidirectional antennas wastes a large portion of the network capacity by reserving the wireless media over a large area.

For example, lots of nodes in the neighborhood of S and D have to sit idle, waiting for the data communication between S and D to finish (see Figure 5.2).

It has been shown earlier that the use of directional antennas can largely reduce the radio interference, thereby improving the utilization of the wireless medium and consequently the network throughput [11–15]. For example, when S and D are communicating with directional antennas at S and D directed towards each other (see Figure 5.2), node H can communicate with J with its antenna directed towards J, since this communication will not disturb the reception at D from S.

However, in the context of ad hoc networks, it is difficult to find ways to control the direction of adaptive antennas for transmission and reception in each terminal in order to achieve an effective multihop communication between any source and destination. This difficulty is mainly due to mobility and lack of centralized control in ad hoc networks. Thus, developing a suitable MAC and routing protocol in ad hoc network using adaptive antennas is a challenging task.

Some researchers in the past have tried to address this challenge in several ways. For example, it has been proposed in [12] to use directional antennas for performance improvement in slotted ALOHA multihop packet radio networks. MAC protocols using directional antennas have also been proposed in [11], where each station is assigned a tone, which is unique to its neighbors. When a station receives a packet, it broadcasts its tone immediately for a period of time so that its neighbor can identify its presence and avoid transmitting in its direction.

In order to tackle the hidden terminal problems, two MAC protocols that rely on RTS/CTS-type handshaking, as in IEEE 802.11, have been suggested in the recent past in the context of ad hoc networks with directional antennas [13, 14]. In [13], a set of directional medium access control (D-MAC) schemes was proposed to show performance improvement over omnidirectional MAC as in IEEE 802.11. Here the mobile nodes are assumed to know the physical locations of themselves and their neighbors using GPS. In [14] the proposed MAC protocol need not know the location information; the source and destination nodes identify each other's direction during RTS/CTS exchange. It is assumed that all the neighbors of source and destination who hear this RTS/CTS dia log will use this information to prevent interfering with the ongoing data transmission.

In order to fully exploit the capability of directional antennas, whenever a source S and destination D engage in a communication, all the neighbors of source and destination nodes should know the direction of communication so that they can initiate new communication in other directions, thus preventing interference with ongoing data communication between S and D. This has been achieved in [11] by using a set of tones and maintaining extensive network status information at each node in the network. In [14] it is assumed that all nodes are able to maintain a unified and common coordinate system to mark the orientation of antennas with respect to each other at all times, irrespective of their movements. They suggest using some direction-finding instrument such as a compass in each node, and this requires additional hardware in each user terminal. Moreover, the probability of control packet collisions [13] has not been taken into account in [14]. So, an appropriate null-steering mechanism needs to be implemented to avoid control packet collisions and to increase the system throughput through overlapping communication.

An adaptive MAC protocol based on electronically steerable passive array radiator (ESPAR) antennas has been proposed [15] to overcome these difficulties. Each node keeps certain neighborhood information dynamically through the maintenance of an Angle–Signal-to-Interference-and-Noise Ratio (SINR) Table (AST), so that each node knows the direction of communication events going on in its neighborhood at that instant of time. Moreover, appropriate mechanisms for null steering of directional antennas in user terminals can help exchange the neighborhood information in the presence of an ongoing communication and can drastically improve the medium utilization through overlapping communications in different directions. The AST will also improve the performance of directional routing, since it helps each node determine the best possible direction of communication with any of its neighbors.

In order to make the directional routing effective, a node should know how to effectively set its transmission direction to transmit a packet to its neighbors. So, each node periodically collects its neighborhood information and forms an AST. $SINR^u_{n,m}(t)$ is a number associated with each link $l^u_{n,m}$, and is a measurable indicator of the strength of radio connection from node n to m at an angle u with respect to n and as perceived by m at any point of time t. The AST of node n specifies the strength of radio connection of its neighbors with respect to n at a particular direction.

Affinity of node m with respect to node n, $a^w_{n,m}(t)$, is a number associated with a link $l^w_{n,m}$ at time t, such that $a^w_{n,m}(t) = \text{Max } [SINR^u_{n,m}(t), 0 < u < 360]$. In other words, the transmission angle w with respect to n maximizes the strength of radio connection from n to m, as perceived by m at any point of time. This maximum SINR value is the affinity of m with respect to n, and this is obtainable when the antenna at n is directed towards m at an angle w with respect to n. Based on this, a Neighborhood-Link-State Table (NLST) at each node is formed.

In this MAC protocol, initially, when node n wants to communicate with m, it sends omnidirectional RTS to inform all the neighbors of n, including m, that a communication from n to m has been requested. It also specifies the approximate duration of communication between n and m. All the neighboring nodes of n keep track of this request from node n, whose direction is known to each of them from the respective AST. The target node *m* sends omnidirectional CTS to grant the request and to inform the neighbors of *m* that *m* is receiving data from *n*. It also specifies the approximate duration of communication. All the neighboring nodes of *m* keep track of the receiving node *m*, whose direction is known to each of them from the respective AST.

On receiving CTS, node *n* issues omnidirectional start-of-data communication (SDC) to inform that the data communication will start from *n* to *m*. If, after getting RTS, SDC is not received within a timeout, RTS is ignored. The receiver acknowledges completion of a successful data communication by sending an ACK to the transmitter.

Other nodes in the neighborhood of *n* and *m* can issue both RTS and CTS without disturbing the communication between *n* and *m*, which is illustrated below. Let us assume another pair of nodes *X* and *Y*, both in the neighborhood of *n* and *m*, desires to communicate. Both of them have already received RTS/CTS from *n* and *m*. From their respective ASTs, both *X* and *Y* know the direction of *n* and *m*. If the directional beam from *X* to *Y* captures *n* or *m*, then the node *X* has to sit idle and defer its desire. Otherwise, node *X* can issue an RTS. In other words, *X* can issue an RTS only if this communication does not interfere with *n* or *m*. The RTS issued by *X*, however, will be selectively omnidirectional: *X* will issue the RTS avoiding interference with *n* and *m*. Similarly, *Y* will respond to this RTS by sending a CTS if the directional beam from *Y* to *X* does not captures *n* or *m*. The CTS issued by *Y* will also be selectively omnidirectional: *Y* will issue a CTS avoiding interference with *n* and *m*.

Some nodes around *n* and *m* will not receive RTS/CTS from *X* and *Y* and therefore will be unaware of this communication event between *X* and *Y*. So, some of these nodes (including *n* or *m*, after the communication between *n* and *m* is over) may initiate another communication, which may disturb the communication between *X* and *Y*. To avoid this, both of the antennas in *X* and *Y* will steer their nulls in the directions towards *n* and *m* so that they will be unaffected by the communication situation described above.

Any other nodes within the transmission beam of an ongoing communication will sit idle during the communication process. But, each of them will be waiting in omnidirectional receive mode with its null steer towards the direction of communication. This will enable the idle nodes to receive RTS/CTS exchange from nodes that are unaware of the communication process. This will happen in the following scenario: When *X* and *Y* are communicating simultaneously with *n* and *m*, some nodes around *n* and *m* will not receive RTS or CTS or both from *X* and *Y* (as mentioned earlier). So, these nodes will be unaware of this communication between *X* and *Y*. So, some of these nodes may initiate another communication. But, they will issue omnidirectional RTS/CTS, since they are unaware of any communication process. All nodes in the *X-Y* beam (excluding *X* and *Y*) need to receive this RTS/CTS to become aware of this new communication. So, during the communication between X and Y, all other nodes in this region will steer their nulls towards X-Y and wait in the omnidirectional receive mode. Nodes X and Y, as mentioned earlier, will steer their nulls in the directions of n and m so that they will be unaffected by the probable communication in those regions. After the communication between X and Y is over, X and Y will collect this information from their neighbors during the next cycle of angle-SINR formation of X and Y.

This can be extended to ensure conflict-free multiple communication simultaneously. The number of simultaneous communications, however, depends on the null-steering capability and also on the current topology of the network.

6.3   Discussions

Researchers are still working to address several performance concerns of MAC protocols proposed in the context of ad hoc networks [16]. One of the important issues is to ensure fairness in medium access. In a multihop configuration using CSMA/CA-based MAC protocols with exponential back-off mechanism, all the nodes may not be able to access the shared medium equally well all of the time. This results in severe performance degradation. The performance degrades further if we assume asymmetric links. Note that all of the above protocols assume the presence of symmetric links. This is valid for a network in which all nodes transmit at the same power level. However, an ad hoc network may be comprised of devices that have different transmit power capabilities. In any event, it will be critical to ensure that the MAC protocol in use does not unduly favor devices that can transmit at higher power levels [17].

Researches are also focusing on power-controlled MAC protocols that would improve channel utilization to a large extent [18, 19]. The basic idea behind conventional CSMA/CA MAC protocol is to reserve the transmission and reception areas of both source and destination. The goal of power-controlled MAC is that, a pair of communicating nodes adjusts their power level to acquire the minimum area of the floor that is needed for it to successfully complete a data transmission [19]. Power-controlled multiple access is achieved by adhering to two key principles [19]:

1. The power conserving principle dictates that each station must transmit at the minimum power level that is required to be successfully heard by its intended receiver.

2. The cooperation principle dictates that no station that commences a new transmission can transmit loud enough to disrupt ongoing transmissions.

Enforcing these two principles can help achieve efficient power-controlled multiple accesses within the framework of collision avoidance protocols.

References

[1] Abramson, N., "The ALOHA System – Another Alternative for Computer Communications," *Proc. Fall Joint Comput. Conf., AFIPS Conf.*, 1970, pp. 281–285.

[2] Kleinrock, L., and F. A. Tobagi, "Packet Switching in Radio Channels: Part I—Carrier Sense Multiple-Access Modes and Their Throughput-Delay Characteristics," *IEEE Trans. Commun.*, Vol. COM-23, No. 12, Dec. 1975, pp. 1400–1416.

[3] Tobagi, F. A., and L. Kleinrock, "Packet Switching in Radio Channels: Part II. The Hidden Terminal Problem in Carrier Sense Multiple Access Modes and the Busytone Solution," *IEEE Trans. on Commun.*, Vol. COM-23, No. 12, 1975, pp. 1417–1433.

[4] Karn, P., "MACA—A New Channel Access Method for Packet Radio," *ARRL/CRRL Amateur Radio 9th Computer Networking Conf.*, 1990, pp. 134–140.

[5] Bharghavan, V., et al., "MACAW: A Media Access Protocol for Wireless LAN's," *ACM SIGCOMM'94*, 1994, pp. 212–225.

[6] Fuller, L., and J. J. Garcia-Luna-Aceves, "Floor Acquisition Multiple Access (FAMA) for Packet-Radio Networks," *ACM SIGCOMM'95*, 1995, pp. 262–273.

[7] Deng, J., and Z. J. Haas, "Dual Busy Tone Multiple Access (DBTMA): A Medium Access Control for Multihop Networks," *Proc. of the IEEE Wireless Communications and Networking Conf.*, New Orleans, LA, Sept. 1999.

[8] Tobagi, F. A., and L. Kleinrock, "Packet Switching in Radio Channels: Part III. Polling and (Dynamic) Split-Channel Reservation Multiple Access Solution," *IEEE Trans. on Commun.*, Vol. COM-24, No. 8, 1976, pp. 832–845.

[9] Lough, Daniel L., T. Keith Blankenship, and Kevin J. Krizman, *A Short Tutorial on Wireless LANs and IEEE 802.11*, The Bradley Department of Electrical and Computer Engineering, Virginia Polytechnic Institute and State University, Blacksburg, VA, Apr. 2000.

[10] Ng, Mario Joa, "Routing Protocol and Medium Access Protocol for Mobile Ad Hoc Networks," Ph.D. Dissertation, Polytechnic University, Jan. 1999.

[11] Yum, T. S., and K. W. Hung, "Design Algorithms for Multihop Packet Radio Networks with Multiple Directional Antennas Stations," *IEEE Trans. on Commun.*, Vol. 40, No. 11, 1992, pp. 1716–1724.

[12] Zander, J., "Slotted ALOHA Multihop Packet Radio Networks with Directional Antennas," *Electronics Letters*, Vol. 26, No. 25, 1990.

[13] Ko, Y. B., V. Shankarkumar, and N. H. Vaidya, "Medium Access Control Protocols Using Directional Antennas in Ad Hoc Networks," *Proc. of IEEE INFOCOM 2000*, March 2000.

[14] Nasipuri, A., et al., "A MAC Protocol for Mobile Ad Hoc Networks Using Directional Antennas," *Proc. of the IEEE WCNC 2000*, 2000.

[15]   Bandyopadhyay, S., et al., "An Adaptive MAC Protocol for Wireless Ad Hoc Community Network (WACNet) Using Electronically Steerable Passive Array Radiator Antenna," *GLOBECOM 2001*, San Antonio, TX, Nov. 2001.

[16]   Xu, S., and T. Saadawi, "Does the IEEE 802.11 MAC Protocol Work Well in Multihop Wireless Ad Hoc Networks?" *IEEE Communication Magazine*, June 2001, pp. 130–137.

[17]   Poojary, N., S. V. Krishnamurthy, and S. Dao, "Medium Access Control in a Network of Ad Hoc Nodes with Heterogeneous Transmit Power Capabilities," *Proceedings of ICC 2001*, Helisinki, 2001.

[18]   Wu, Shih-Lin, and Yu-Chee Tseng, "Intelligent Medium Access for Mobile Ad Hoc Networks with Busy Tones and Power Control," *JSAC*, Vol. 18, No. 9, Sept. 2000.

[19]   Monks, Jeffrey P., Vaduvur Bharghavan, and Wen-mei Hwu, "A Power-Controlled Multiple Access Protocol for Wireless Packet Networks," *IEEE INFOCOM 2001*, Anchorage, Alaska, Apr. 2001.

7

Routing Protocols in Ad Hoc Wireless Networks

In this chapter, we will discuss various routing protocols proposed so far in the context of ad hoc networks. Both unicast and multicast routing protocols will be examined and the performance of some of the major unicast and multicast routing protocols will be illustrated.

7.1 Introduction

The problem of routing in a network has two components: *route discovery* and *route maintenance*. In order to communicate with a destination, a source needs to discover a suitable route for sending packets to that destination. However, as the status of different links (or routers) used in this route change, changes in the route may be necessary and a new route may need to be discovered. Conventional routing protocols for wired networks commonly use either *distance vector* or *link-state* protocols [1]. These protocols integrate route discovery with route maintenance by sending routing update packets. If the status of a link or router changes, the update packets will eventually reflect the changes to all other routers. As a result, new routes may be computed, if needed. These protocols can be termed *proactive protocols* because they attempt to discover and maintain routes continuously so that routes are known a priori when needed.

In *distance vector routing* (DVR) protocols, each router periodically broadcasts to its neighbors its view of the distance to all other routers, and each router computes the shortest path to every other router based on the advertised information. By comparing the distances received for each destination from each of its neighbors, a router can determine which of its neighbors is the correct "next hop" on the shortest path toward each destination. When a source wants to send a data packet to a destination, the source simply forwards the data packet to the correct next-hop router. This router, in turn, forwards the packet to the correct next hop. This will continue till the packet reaches its destination. DVR suffers from slow convergence and loop formation. These loops are formed because each router chooses its next hops in a distributed fashion based on possibly old and invalid information. Furthermore, routing updates may have to be propagated to a large number of routers. Examples of DVR protocols include the routing protocol used in the DARPA Packet Radio Network [2] and the original routing protocol for the ARPANET [3].

In *link-state routing* (LSR) protocols, each router maintains its view of the entire network topology. To maintain this, each router periodically broadcasts a link-state packet (LSP) containing a list of its neighbors and the cost of each link to its neighbors. This information from each router eventually reaches all other routers in the network in a periodic fashion. Given this information about the cost of links in the network, each router can compute the best path to each possible destination. When a source wants to send a data packet to a destination, the source computes the best path to the destination and then forwards the data packet to the correct next hop router based on this computed best path. This router, in turn, computes the best path towards the destination and forwards the packet to the correct next hop. This will continue until the packet reaches its destination. LSR protocols converge much more quickly when conditions in the network change. However, it generally requires more computation time than that taken by the distance vector algorithm because each router computes the complete shortest path to each possible destination. It also consumes more network bandwidth than the distance vector algorithm because the routing updates from each router are broadcast to all other routers in the entire network. Examples of LSR protocols include the "new" routing protocol that replaced the original protocol for the ARPANET [3] and open shortest path first (OSPF) [4].

A natural method for trying to provide routing in an ad hoc network is to simply treat each mobile host as a router and to run a conventional routing protocol between them [2]. However, use of these conventional routing protocols in an ad hoc network, treating each mobile host as a router, gives rise to a series of problems, as illustrated in [5, 6]:

- Transmission between two hosts over a wireless network does not necessarily work equally well in both directions. Thus, some routes determined by conventional routing protocols may not work in some environments.

- Wired networks are usually explicitly configured to have only a small number of routers connecting any two networks. In the ad hoc wireless environment, each node is a router and too many paths exist between any two nodes in the network. These redundant paths unnecessarily increase the size of routing updates that must be sent over the network and increase the computational as well as communication overhead.

- Periodic broadcasting of routing updates wastes network bandwidth in ad hoc wireless environments. Routing updates from mobile hosts outside each other's transmission range will not interfere with each other, but where many mobile hosts are within transmission range of each other, their routing updates will consume each other's network bandwidth.

- In a highly dynamic ad hoc network, the rate of topology change may be higher than the rate of route requests. In such an environment, periodic updates could be a waste because most of this routing update information is never used.

- Most mobile hosts in an ad hoc network will be operating on a limited battery power, and periodically sending routing updates wastes battery power.

So we find that the conventional, proactive routing protocols place too heavy a computation and communication demand on each mobile host in multihop wireless networks. Additionally, convergence characteristics of these protocols are not good enough for ad hoc networks [6]. The speed of convergence may be improved by sending routing updates more frequently, but this will waste more bandwidth and battery power. Moreover, this may be a wasteful effort altogether when topology changes are less frequent. Keeping these factors in mind, several researchers have proposed modifications of these conventional protocols to suit the environment of ad hoc wireless networks. For example, researchers have focused on restricting the propagation of routing updates, thereby reducing the control overheads [7]. At the same time, a lot of work has been done in designing and developing alternative demand-driven routing protocols [8], termed *reactive protocols*. These protocols are based on separate route discovery and route maintenance protocols and do not require periodic route updates; they invoke route determination procedure only on demand. Whenever a source node requires a route to a destination, it initiates a route-discovery process within the network. The source node broadcasts a route-request packet to its neighbors, which in turn forward the request to their neighbors, and so on, until the destination is located. The destination then responds by sending a route-reply packet back to the source node through the intermediate nodes. Once a route has been established, it is maintained by a route-maintenance procedure until either the destination becomes inaccessible from the source, or the route is no longer desired.

The performance of the above approach depends on two major factors: (1) how frequently a route discovery is needed, and (2) how often route maintenance is needed. If the communication requirement is high, frequent route discovery would be needed. Since each route discovery is associated with flooding of route-request packets throughout the network, multiple route discoveries at a time will flood the network with route-request packets, increasing congestion and collision in the network. At the same time, if the nodes are mobile, requiring frequent route maintenance, this would in turn increase the control packets in the network further. In the reactive mechanism, computation and communication overhead are reduced only when the demand for communication is less. Moreover, this reduced computation and communication overhead is achieved at the expense of a longer delay in route-setup procedure. Performance can be improved through aggressive route caching and full use of information in the cache [6]. However, additional effort would be needed to detect the staleness of the information in the cache. Another way to decrease control packets is directional propagation of route-request packets. All of these issues will be discussed in detail in Section 7.2.2.

7.2 Unicast Routing Protocols in Ad Hoc Networks

7.2.1 Proactive Routing Protocols

7.2.1.1 Destination-Sequenced Distance Vector Routing Protocol (DSDV)

In DSDV [9], each node maintains a routing table, which has an entry for each destination in the network. The attributes for each destination are the next hop, metric (hop counts), and a sequence number originated by the destination node. To maintain the consistency of the routing tables, DSDV uses both periodic and triggered routing updates; triggered routing updates are used in addition to the periodic updates in order to propagate the routing information as quickly as possible when there is any topological change. The update packets include the destinations accessible from each node and the number of hops required to reach each destination along with the sequence number associated with each route.

Upon receiving a route-update packet, each node compares it to the existing information regarding the route. Routes with old sequence numbers are simply discarded. In case of routes with equal sequence numbers, the advertised route replaces the old one if it has a better metric. The metric is then incremented by one hop since incoming packets will require one more hop to reach the destination. A newly recorded route is immediately advertised to its neighbors.

When a link to the next hop is broken, any route through that next hop is immediately assigned an infinity metric and assigned an updated sequence number. This is the only case when sequence numbers are not assigned by the destination. When a node receives an infinity metric and it has an equal or later sequence number with a finite metric, a route update broadcast is triggered. Therefore, routes with infinity metrics will be quickly replaced by real routes propagated from the newly located destination.

One of the major advantages of DSDV is that it provides loop-free routes at all instants. It has a number of drawbacks however. Optimal values for the parameters, such as maximum settling time, for a particular destination are difficult to determine. This might lead to route fluctuations and spurious advertisements resulting in waste of bandwidth. DSDV also uses both periodic and triggered routing updates, which could cause excessive communication overhead. In addition, in DSDV, a node has to wait until it receives the next route update originated by the destination before it can update its routing table entry for that destination. Furthermore, DSDV does not support multipath routing.

7.2.1.2   Wireless Routing Protocol (WRP)

WRP [10] is also based on distance vector algorithm. To avoid the looping problem present in the distance vector algorithm, WRP includes second-to-last hop (predecessor) information for each destination. This predecessor information is included in two tables kept at each node: a distance table and a routing table. The distance table of node i is a matrix containing, for each destination j and each neighbor k, the distance to j and the predecessor when k is chosen as the next hop to reach j. The routing table contains an entry for each destination, and the attributes for each destination include the next hop, distance, and the predecessor.

To keep a consistent view of the network and to respond to topological changes, each node exchanges update packets with its neighbors. An update entry specifies a destination, a distance to the destination, and a predecessor to the destination. Upon receiving an update packet, each node updates the distance and the predecessor entries in the distance table. Let's say a node i receives an update from its neighbor k regarding destination j. In addition to updating the distance and the predecessor information for the destination j and neighbor k pair, WRP determines if the path to destination j through any other neighbors includes node k and updates affected entries. Note that by back-tracing the predecessor information kept in the distance table, the complete path can be found. After updating the distance table, each node also updates its routing table accordingly by choosing the neighbor that offers the smallest cost to the given destination as the next hop to reach that destination.

The major advantage of WRP is that it reduces temporary looping by using the predecessor information to identify the route. This in turn reduces the convergence time. However, as in all the proactive protocols, in WRP, each node constantly maintains full routing information to all destinations in the network, requiring significant communication overhead.

7.2.1.3   Cluster-Based Routing (CBR)

In CBR protocol [11], a network is divided into a number of overlapping clusters whose union covers the entire network. This is shown in Figure 7.1. A cluster consists of nodes, which are reachable in one hop, and there is one boundary node between overlapping clusters. In CBR, each node maintains a list of its neighbors, a list of clusters in the network, and a list of boundary nodes in the network. Based on this information, each node can keep a view of the entire network topology. For a connected network, the boundary nodes form a connected network. To construct routes, a shortest path algorithm is applied on this connected network, which consists of boundary nodes.

In CBR, a topological change corresponds to a change in cluster membership. Upon receiving new cluster information, a boundary node updates its cluster list and boundary node list, and broadcasts this information. Nonboundary nodes just update their lists and do not broadcast. Since only the boundary nodes are responsible for broadcasting any new information, the number of topology update packets is reduced. This is the major advantage of CBR. The cost of this method, however, is formation and maintenance of clusters. For example, when a new host comes up, the protocol needs to create new clusters, determine which clusters are absolutely essential, and remove all the redundant clusters, since the network needs to remain cluster connected with the optimum number of clusters. In addition, temporary routing loops may exist in the network since some nodes may have inconsistent topology information due to long propagation delays. CBR also does not guarantee the shortest path since the first-fit approach is taken when the clusters are created.

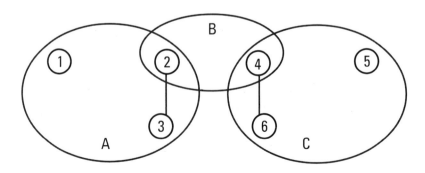

Figure 7.1   Formation of overlapping clusters in CBR.

7.2.1.4   Cluster Head-Gateway Switch Routing (CGSR)

CGSR protocol [12] is also based on a cluster-based network architecture. In CGSR, specific nodes are designated to be cluster heads, which control channel access within a cluster at MAC layer. At MAC layer, cluster heads are given a priority, and thus they have more chances to transmit than other nodes. All nodes within a cluster can communicate with its cluster head and possibly with each other. To prevent frequent cluster head changes, the Least Cluster Head Change Algorithm (LLC) is used.

In CGSR, the DSDV protocol is modified so that it can take an advantage of the clustering architecture. Specifically, hierarchical routing is used to route packets. Each node maintains two tables: a cluster member table which maps the destination node address to its cluster head address, and a routing table which shows the next hop to reach the destination cluster. Both tables contain sequence numbers to purge stale routes and prevent looping. In CGSR, packets are routed alternatively through cluster heads and gateways. In other words, the typical route looks like $C_1G_1C_2G_2...C_lG_l$, where C_i is a cluster head and G_i is a gateway. This is shown in Figure 5.1 in Chapter 5. A gateway refers to a node that belongs to more than one cluster. Therefore, in CGSR, gateway-to-gateway shortcuts are avoided. Even though this strict alternate routing results in increased path length, this drawback is outweighed by its merits. Specifically, the presence of a cluster head between two gateways is advantageous since cluster heads have more chances to transmit than other nodes and a direct gateway-to-gateway communication requires that both gateways meet on the same code for transmission, which can lead to code scheduling overhead and hence delay.

The major advantage of CGSR is that only the routes to the cluster heads are maintained due to the hierarchical routing. However, there is overhead associated with maintaining clusters. Specifically, each node needs to periodically broadcast its cluster member table and update its table based on the received updates.

7.2.1.5   Fisheye State Routing

Fisheye State Routing (FSR) [7] introduces the notion of multilevel fisheye scope to reduce routing update overhead in large networks. Nodes exchange link-state entries with their neighbors with a frequency that depends on distance to destination. From link-state entries, nodes construct the topology map of the entire network and compute optimal routes. FSR uses the "fisheye" technique proposed by Kleinrock and Stevens [13], where the technique was used to reduce the size of information required to represent graphical data. The eye of a fish captures with high detail the pixels near the focal point. The detail decreases as the distance from the focal point increases. In routing, the fisheye approach translates to maintaining accurate distance and path quality information about the immediate neighborhood of a node, with progressively less detail as the distance increases. FSR is functionally similar to LS Routing in that it maintains a topology map at each node. The key difference is the way in which routing information is disseminated. In LS, link-state packets are generated and flooded into the network whenever a node detects a topology change.

In FSR, link-state packets are not flooded. Instead, nodes maintain a link-state table based on the up-to-date information received from neighboring nodes, and periodically exchange it with their local neighbors only (no flooding). Through this exchange process, the table entries with larger sequence numbers replace the ones with smaller sequence numbers. The FSR periodic table exchange resembles the vector exchange in DSDV [9] where the distances are updated according to the time stamp or sequence number assigned by the node originating the update. However, in FSR link states rather than distance vectors are propagated. Moreover, like in LS, a full topology map is kept at each node and shortest paths are computed using this map.

In order to reduce the size of update messages without seriously affecting routing accuracy, FSR uses the notion of multilevel fisheye scope. The scope is defined as the set of nodes that can be reached within a given number of hops. The reduction of routing update overhead is obtained by using different exchange periods for different entries in routing table. More precisely, entries corresponding to nodes within the smaller scope are propagated to the neighbors with the highest frequency. The rest of the entries are sent out at a lower frequency. As a result, a considerable fraction of link-state entries are suppressed in a typical update, thus reducing the message size. This strategy produces timely updates from near stations, but creates large latencies from stations afar. However the imprecise knowledge of the best path to a distant destination is compensated by the fact that the route becomes progressively more accurate as the packet gets closer to destination. As the network size grows large, a "graded" frequency update plan must be used across multiple scopes to keep the overhead low.

7.2.2   Reactive Routing Protocols

7.2.2.1   Dynamic Source Routing (DSR)

Dynamic source routing [14] is based on source routing, where the source specifies the complete path to the destination in the packet header and each node along this path simply forwards the packet to the next hop indicated in the path. It utilizes a route cache where routes it has learned so far are cached. Therefore, a source first checks its route cache to determine the route to the destination. If a route is found, the source uses this route. Otherwise, the source uses a route discovery protocol to discover a route.

In route discovery, the source floods a query packet through the ad hoc network, and the reply is returned by either the destination or another host, which can complete the query from its route cache. Each query packet has a unique ID and an initially empty list. When receiving a query packet, if a node has already seen this ID (i.e., duplicate) or it finds its own address already recorded in the list, it discards the copy and stops flooding; otherwise, it appends its own address in the list and broadcasts the query to its neighbors. If a node can complete the query from its route cache, it may send a reply packet to the source without propagating the query packet further. Furthermore, any node participating in route discovery can learn routes from passing data packets and gather this routing information into its route cache.

A route failure can be detected by the link-level protocol (i.e., hop-by-hop acknowledgments), or it may be inferred when no broadcasts have been received for a while from a former neighbor. When a route failure is detected, the node detecting the failure sends an error packet to the source, which then uses route discovery protocol again to discover a new route. Note that in DSR, no periodic control messages are used for route maintenance.

The major advantage of DSR is that there is little or no routing overhead when a single or few sources communicate with infrequently accessed destinations. In such situation, it does not make sense to maintain routes from all sources to such destinations. Furthermore, since communication is assumed to be infrequent, a lot of topological changes may occur without triggering new route discoveries.

Even though DSR is suitable for the environment where only a few source communicate with infrequently accessed destinations, it may result in large delay and large communication overhead in highly dynamic environment with frequent communication requirement [15]. Furthermore, DSR may have scalability problem [16]. As the network becomes larger, control packets and message packets also become larger since they need to carry addresses for every nodes in the path. This may be a problem since ad hoc networks have limited available bandwidth.

7.2.2.2   Ad Hoc On-Demand Distance Vector Routing (AODV)

The Ad Hoc On-Demand Distance Vector (AODV) [17, 18] routing protocol shares DSR's on-demand characteristics in that it also discovers routes on an "as needed" basis via a similar route discovery process. However, AODV adopts a very different mechanism to maintain routing information. It uses traditional routing tables, one entry per destination. This is a departure from DSR, which can maintain multiple route cache entries for each destination. Without source routing, AODV relies on routing table entries to propagate a route reply back to the source and, subsequently, to route data packets to the destination. AODV uses sequence numbers maintained at each destination to determine freshness of routing information and to prevent routing loops. These sequence numbers are carried by all routing packets.

When a route is needed, a node broadcasts a route request message. The response message is then echoed back once the request message reaches the destination or an intermediate node that contains a fresh route to the destination. For each route, a node also maintains a list of those neighbors actively using the route. A link breakage causes immediate link failure notifications to be sent to the affected neighbors. Similar to DSDV, each route table entry is tagged with a destination sequence number to avoid loop formation. Moreover, nodes are not required to maintain routes that are not active. Thus, wireless resource can be effectively utilized. However, since flooding is used for route search, communication overhead for route search is not scalable for large networks. As route maintenance considers only the link breakage and ignores the link creation, the route may become nonoptimal when network topology changes. Subsequent global route search is needed when the route is broken.

An important feature of AODV is maintenance of timer-based states in each node, regarding utilization of individual routing table entries. A routing table entry is "expired" if not used recently. A set of predecessor nodes is maintained for each routing table entry, indicating the set of neighboring nodes that use that entry to route data packets. These nodes are notified with route error packets when the next hop link breaks. Each predecessor node, in turn, forwards the route error to its own set of predecessors, thus effectively erasing all routes using the broken link.

The recent specification of AODV [19] includes an optimization technique to control the route request flood in the route discovery process. It uses an expanding ring search initially to discover routes to an unknown destination. In the expanding ring search, increasingly larger neighborhoods are searched to find the destination. The search is controlled by the TTL field in the IP header of the route request packets. If the route to a previously known destination is needed, the prior hopwise distance is used to optimize the search.

7.2.2.3 Temporally Ordered Routing Algorithm

Temporally Ordered Routing Algorithm (TORA) [16] is a highly adaptive loop-free distributed routing protocol for multihop networks. A key concept in the protocol's design is an attempt to decouple the generation of far-reaching control message propagation from the dynamics of the network topology. The basic, underlying algorithm is one of a family of algorithms referred to as *link reversal* algorithms. In particular, the protocol's reaction to certain link failures is structured as a temporally ordered sequence of diffusing computations, each computation consisting of a sequence of directed link reversals.

TORA is partly based on the classical work by Gafni and Bertsekas [20] in order to maintain a destination-oriented directed acyclic graph (DAG) in the face of topology changes. A DAG is considered destination oriented, if for every node there is a path to a given destination. If this configuration is lost due to link failures, a series of link reversals take place to ensure that the graph again becomes destination-oriented in finite time. Based on the query/reply process proposed in [21], a sequence of directed links leading from the source to the destination is formed (route discovery and construction phase) in a source-initiated fashion. From that point it is maintained (route maintenance phase) using link reversal, whenever topology change causes a node to loose its last downstream link. If the destination becomes unreachable because of a network partition, the protocol erases (route eraser phase) all invalid routes.

TORA provides multiple paths to a destination and ensures they are loop-free. This is a destination-oriented protocol in which logically separate version of the protocol is run for each destination in active communication. However, after the DAG creation, new links would not be considered unless the DAG becomes disconnected. Therefore, the route may become nonoptimal. In addition, communication overhead for route creation is not scalable because flooding is used. TORA does not guarantee generation of shortest path routes. Another drawback of TORA is that it may take long time for the protocol to converge if a link failure partitions the network.

7.2.2.4 Associativity-Based Routing (ABR)

Associativity-based routing [22] also uses a source-initiated method, that is, ABR only maintains routes for sources that actually desire routes. To find a route to the desired destination, ABR also uses query-reply control packets. As in TORA and DSR, the source floods a query packet through the ad hoc network to discover a route. However, unlike in TORA and DSR, the best route is selected by the destination based on the stability of the route and shortest path. A route is considered to be stable if it consists of nodes that have been stationary for a threshold period. The stationary period is measured in terms of associativity ticks, which are described below.

Each node maintains multiple associativity ticks (one per each neighbor), which are initially set to zero. Periodically, each node broadcasts beacons identifying itself. Note that these beacons are not newly defined messages, but are link-level messages, which are normally exchanged between nodes to maintain connectivity. Therefore, no extra overhead is added. Each time a beacon is received, a node increments the associativity tick which is associated with the node from which a beacon is received. If a neighboring node moves out of proximity, the associativity tick, which is associated with that neighbor is reset to zero. Therefore, if a mobile host has low associativity ticks with its neighbors, it indicates that the mobile host is in high state of mobility. On the other hand, if a mobile host has high associativity ticks with its neighbors, it indicates that the mobile host is in the stable (or stationary) state.

In route discovery, the source floods a query packet. As in DSR, a query packet contains an initially empty list, and when receiving a query packet, a node appends its own address in the list. In ABR, however, a query packet also contains the associativity ticks along with other metrics such as hop count. When a node receives a query packet, it appends its address in the list, and it also appends the associativity ticks with its neighbors along with other metrics. The next succeeding node keeps only its upstream neighbor's associativity tick, which is associated with it and erases all others. Therefore, when the query packet reaches the destination, it will contain only the intermediate nodes' addresses and the associativity ticks for that path, together with routing information such as hop count.

After some time period, the destination will know all the possible routes and their qualities. It then selects the best route based on the selection criteria. The most important criterion in selecting a route is the stability of the route indicated by the associativity ticks. Therefore, the destination chooses the route that consists of nodes having high associativity ticks. If more than one route has the same degree of stability, the destination selects the shortest route. The destination then sends a reply packet back to the source. Only the best route selected will be valid, while all other routes will be inactive. When a node moves, unless it is the source, its immediate upstream node erases its route and tries to find an alternate route using the localized query, thereby retaining the control packets locally. Only when necessary, the full query-reply process is performed. Source movement always invokes the full query-reply process.

The major advantage of ABR is that routes selected tend to be long-lived since nodes that have been stationary for some time are less likely to move. This results in fewer route reconstructions, thereby reducing the communication overhead. A possible drawback of ABR is that ABR may have scalability problem due to the limited available bandwidth, since control packets and message packets need to carry addresses for every nodes in the path as in DSR.

7.2.2.5   Signal Stability-Based Adaptive Routing (SSAR)

SSAR [23] also discovers a route only when desired by a source. To find a route to the desired destination, SSAR also uses a query-reply process; the source floods a query packet, and the destination determines the best route and returns a reply packet to the source. The novelty of the SSAR protocol is the use of signal strength as a route-selection criteria. The basic idea of SSAR is to select the route that consists of links with the strong signals since such routes lead to longer-lived routes due to the "buffer zone" effect and consequently require less route maintenance. The buffer zone effect is as follows: if a link between two nodes is a strong link (i.e., a link with strong signal strength), it will have to become a weak link before triggering a route reconstruction. In other words, two nodes that make up the strong link can roam within a certain vicinity of each other without triggering a route reconstruction.

To take the signal strength into account in routing process, each node periodically broadcasts link-level beacons identifying itself, and upon receiving a beacon, each node records the signal strength at which the beacon was received. If the signal strength is above the threshold, the link is marked as a strong link. Otherwise, it is marked as a weak link. When a node receives a query packet, it propagates the query packet further only if the query packet is received over a strong link and the node has not seen this query packet before. A query packet that is received over a weak link is dropped. When a query packet reaches the destination, it contains the address of each intermediate node. The destination selects the route recorded in the first received query packet since it is probably the shortest path. It then sends a reply packet back to the source along the selected route.

Since only the query packets that are received over the strong links are forwarded further, a query packet may never reach the destination if there is no route which consists of strong links. In this case, the source will not receive any reply packet within some timeout period. When the timeout period expires, the source can wait and try again later, or it may wish to find any route by specially marking the query packet. When a route failure is detected, the node detecting the failure sends an error packet to the source, which then sends a message to erase the invalid route and uses route discovery protocol again to find a new route.

The major advantage of SSAR is that routes selected tend to be long-lived, and this in turn reduces the number of route reconstructions required. Furthermore, SSAR results in low packet loss since strong links are less vulnerable to interference than weak links. A possible drawback of SSAR is that SSAR results in routes with slightly higher hop counts than optimal routing since SSAR prefers routes with strong links, which are likely to be between two hosts close to each other.

7.2.2.6 Stability-Based Routing

In an ad hoc network, relationship among nodes is based on providing some kind of service, and stability can be defined as the minimal interruption in that service. Irrespective of the routing schemes, frequent interruption in a selected route would degrade the performance in terms of quality of service. Therefore, an important issue is to minimize route maintenance by selecting stable routes, rather than shortest route. The notion of stability of a path is dynamic and context-sensitive. Stability of a path is the span of life of that path at a given instant of time and it has to be evaluated in the context of providing a service. A path between a source and destination is said to be stable if its span of life is sufficient to complete a required volume of data transfer from source to destination. Hence, a given path may be sufficiently stable to transfer a small volume of data between source and destination; but the same path may be unstable in a context where a large volume of data needs to be transferred.

In stability based routing [24], a notion of link stability and path stability and their evaluation mechanism in the context of dynamic topology changes in an ad hoc network has been proposed. Subsequently, a distributed routing scheme among mobile hosts is proposed in order to find a path between them that is stable in a specific context.

The strength of relationship between two nodes over a period of time is defined as *node-affinity* or *affinity*. Informally speaking, *link-affinity* $a_{nm}(t)$, associated with a link l_{nm} at time t, is a prediction about the span of life of the link l_{nm} in a particular context. Link-affinity $a_{nm}(t)$ at that instant of time is a function of the current distance between n and m, relative mobility of m with respect to n, and the transmission range of n. If transmission range of n and m are different, $a_{nm}(t)$ may not be equal to $a_{mn}(t)$. The *node-affinity* or *affinity* $\eta_{nm}(t)$ between two nodes and its neighbor m is defined as $\min[a_{nm}(t), a_{mn}(t)]$. The stability of connectivity between n and its neighbor m depends on η_{nm}. The unit of affinity is seconds.

To find out the link-affinity $a_{nm}(t)$ at any instant of time, node n sends a periodic beacon and node m samples the strength of signals received from node n periodically. Since the signal strength of n as perceived by m is a function $f(R_n, d_{nm})$ where R_n is the transmission range of n, and d_{nm} is the current distance between n and m at time t, the node m can predict the current distance d_{nm} at time t between n and m. If M is the average velocity of the nodes, the worst-case link-affinity $a_{nm}(t)$ at time t is $(R_n - d_{nm})/M$, assuming that at time t, the node m has started moving outwards with an average velocity M. For example, If the transmission range of n is 300m, the average velocity is 10 m/s and current distance between n and m is 100m, the life span of link l_{nm} (worst case) is 20s, assuming that the node m is moving away from n in a direction obtained by joining n and m.

The above method is simple, but is based on an optimistic assumption that node-distance can be deduced from signal strength. In real life, even if the transceivers have the same transmission range, it will vary because of the differences in battery power in each of them. Therefore, it may be difficult for one node to estimate distance from another node by monitoring the current signal strength only. So, a node m needs to monitor the change in signal strength of n over time to estimate link-affinity, as described below.

Let $\Delta S_{nm}(t)$ be the change of signal strength at time t and is defined as:

$\Delta S_{nm}(t) = S_{nm}(t) - S_{nm}(t - \Delta t)$, where $S_{nm}(t)$ is the current sample value of the signal strength of node n as perceived by node m at time t, $S_{nm}(t - \Delta t)$ is the previous sample value at time $(t - \Delta t)$ and Δt is the sampling interval. Let $S'_{nm}(t)$ be the rate of change of signal strength at time t and is defined as $S'_{nm}(t) =$

$(\Delta S_{nm}(t) / \Delta t)$ and let $S'_{nm}(t)avg$ is the average rate of change of signal strength at time t over the past few samples. Let S_t be the threshold-signal strength: when the signal strength S_{nm} associated with l_{nm} goes below S_t, we assume that the link l_{nm} is disconnected. Further, we define

$a_{nm}(t)$ = high, if $S'_{nm}(t)$avg is positive;

$\qquad$ = $(S_t - S_{nm}(t)) / S'_{nm}(t)$avg, if $S'_{nm}(t)$avg is negative.

If $\Delta S_{nm(avg)}$ is positive, it indicates that the link-affinity is increasing and the two nodes are coming closer. Hence, link-affinity is termed high at that instant of time. The value for *high* is computed as (transmission range / average node velocity) and is approximately equal to the time taken by a node m to cross the average transmission range of node n with an average velocity. However, as indicated earlier, even if $S'_{nm}(t)$avg is positive, a node $p \in N_n$ in the periphery of the transmission range of n is weakly connected to n compared to a node $p \in N_n$ which is closer to n. Thus, the chance of m going out of the transmission range of n due to a sudden outward mobility of either m or n is more than that of p. Thus, if $S'_{nm}(t)$avg is positive, a correction factor μ is used to moderate this *high value. This correction factor μ is equal to* $(1 - S_t / S_{nm}(t))$ and $aVnm(t) = \mu^*$ *high*, if $S'_{nm}(t)$avg is positive. This indicates that if $S_{nm}(t)$ is very close to S_t, μ will be close to zero and consequently $a_{nm}(t)$ will also become close to zero, even if $S(_{nm}(t)_{!AVG}$ is positive.

As indicated earlier, the *node-affinity* or *affinity* between two nodes n and m, $\eta_{nm}(t)$, is defined as $\min[a_{nm}(t), a_{mn}(t)]$.

Given any path $p = (i, j, k,..., l, m)$, the stability of path p will be determined by the lowest-affinity link (since that is the bottleneck for the path) and is defined as $\min[\eta_{ij}(t), \eta_{jk}k(t),..., \eta_{lm}()t]$. In other words, stability of path p at some instant of time t between source s and destination d, $\eta^p_{sd}(t)$, is given by: $\eta^p_{sd}(t) = min[\eta_{ij}(t)], \forall i, j \in p$.

In Figure 7.2, each link is associated with a value indicating the affinity in seconds between nodes connected by that link. For simplicity, we assume $l_{nm} = l_{mn}$ and $a_{nm}(t) = a_{mn}(t)$. Stability of any path p between C and F, say $CDGF$ is $\min[4, 1, 2]$ (i.e., 1 sec). The stability of shortest path between C and F, CDF, is $\min[4, 0.4]$ (i.e., 0.4 sec).

The basic path-searching mechanism given above is same as in most of the reactive routing with three differences:

- The search is not restricted to find the shortest path; if multiple paths exist between source and destination, the source receives multiple path information from destination in sequence;

- The route reply packet from destination to source would collect the most recent value of affinity a_{mn} for all intermediate nodes m, n, ...;

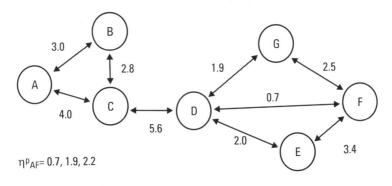

Figure 7.2 An example network with affinity specified.

- Each path between source and destination is associated with a time delay to estimate the delay associated with that path due to traffic congestion.

When a source initiates a route discovery request, it waits for the route reply until timeout. All the route replies received until timeout are cached at the source. Whenever the source receives the first route reply, it knows the path to destination and immediately computes its stability η^p_{sd}. If V is the volume of data in number of packets to be sent to destination and if B is the bandwidth for transmitting data in packets per second, V/B is the one-hop delay to transmit the data, ignoring all other delay factors. If H_p is the number of hops from source to destination in path p, (H_p*V/B) will be the time taken to complete the data transfer. If η^p_{sd} is sufficient to carry this data, the path is selected. Otherwise, the source checks the next path, if available in its cache, for sufficient stability.

When the average mobility of all the nodes in the system is low and/or volume of data to be communicated between source and destination is low, the chance of route error during data communication with shortest-path algorithm would be low. Conventional shortest path routing algorithm would work well in this situation. However, when the average mobility of all the nodes in the system is high and/or volume of data to be communicated between source and destination is high, the chance of route error during data communication with shortest path algorithm would be high. In this situation, we need to find out a stable path rather than shortest path for routing. The results in [24, 25] show that the stable-path algorithm reduces route error drastically in all scenarios.

7.2.2.7 Multipath Routing with Preemptive Route Discovery

As an extension to stability based routing, a multipath routing scheme with preemptive route discovery has been proposed in [26] to reduce end-to-end delay and to ensure better quality of service. The successful operation of an ad hoc network will be disturbed, if an intermediate node, supporting a communication between a source-destination pair, moves out of range during data transfer. This interruption in communication results in subsequent route rediscovery between that source-destination pair or invoking some path-maintenance algorithm that eventually increases the end-to-end delay.

Moreover, the routing schemes proposed so far in the context of ad hoc networks employ single-path routing that might not ensure optimal end-to-end delay. However, once a set of paths between s to d is discovered, in some cases, it is possible to improve end-to-end delay by splitting the volume of data into different blocks and sending it via selected multiple paths from s to d which would eventually reduce congestion and end-to-end delay.

In [26], a mechanism is proposed for adaptive computation of multiple paths in temporal and spatial domain to transmit large volume of data from s to d in ad hoc wireless networks. Two aspects are considered in this framework. The first aspect is to perform preemptive route rediscoveries before the occurrence of route errors while transmitting large volume of data from s to d. Consequently, this helps to find out dynamically a series of multiple paths in the temporal domain to complete the data transfer. The second aspect is to select multiple paths in the spatial domain for data transfer at any instant of time and to distribute the data packets in sequential blocks over those paths in order to reduce congestion and end-to-end delay. A notion of link stability and path stability has been used [24] and a unified mechanism is proposed to address these two aspects that relies on evaluating a path based on link stability and path stability. The solution method uses Lagrangean relaxation and subgradient heuristics [27] to find out the paths and data distribution into those paths both in temporal and spatial domains. The performance of this approach has been evaluated on a simulation environment. It has been observed that the use of temporal multipaths allows any source to transmit a large volume of data to a destination without degradation of performance due to route-errors. Additionally, the use of spatial multipaths would help to significantly reduce the end-to-end delay and the number of route-rediscovery needed in this process.

In Figure 7.3, we have shown the stability of most stable path between two arbitrary nodes, 14 and 5, of an ad hoc network, sampled at 5-second intervals of time. At each 5 seconds, a route discovery process is initiated from node 14 and the paths obtained after route-request–timeout (300 msec) are evaluated to find out the most stable path. As shown in the figure, it has been observed that no single path is stable throughout the span of 30 seconds. However, we are getting a sustained stability between node 14 and 5 through different intermediate nodes. This establishes the viability of our scheme. In other words, if we can perform preemptive route rediscoveries before the occurrence of route errors while transmitting large volume of data from s to d, it is possible to find out

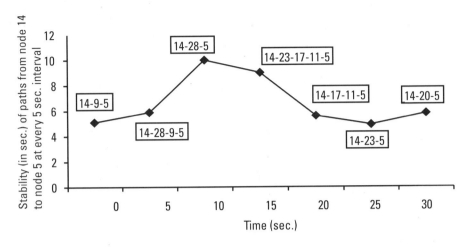

Figure 7.3   Stability of maximum stable path between node 14 and 5, sampled at each 5-second interval (the path is given in the boxes).

dynamically a series of multiple paths in temporal domain to complete a large volume of data transfer.

7.2.2.8   Location-Aided Routing (LAR) Protocols

Location-Aided Routing (LAR) Protocols [28] suggest an approach to decrease overhead of route discovery by utilizing location information for mobile hosts. Such location information may be obtained using the global positioning system (GPS) [29]. The LAR protocols use location information to reduce the search space for a desired route. Limiting the search space results in fewer route discovery messages.

These protocols limit the search for a route to the so-called request zone, determined based on the expected location of the destination node at the time of route discovery. The expected zone of node D, from the viewpoint of node S at time t_1, is the region that node S expects to contain node D at time t_1. Node S can determine the expected zone based on the knowledge that node D was at location L at time t_0. If node S does not know a previous location of node D, then node S cannot reasonably determine the expected zone—in this case, the entire region that may potentially be occupied by the ad hoc network is assumed to be the expected zone. In this case, this algorithm reduces to the basic flooding algorithm. In general, having more information regarding mobility of a destination node, can result in a smaller expected zone.

During route discovery, the source node S defines (implicitly or explicitly) a request zone for the route request. A node forwards a route request only if it belongs to the request zone (unlike the flooding algorithm). To increase the probability that the route request will reach node *D*, the request zone should include the expected zone (described above). Additionally, the request zone may also include other regions around the request zone. Note that the probability of finding a path (in the first attempt) can be increased by increasing the size of the initial request zone. However, route discovery overhead also increases with the size of the request zone. Thus, there exists a trade-off between latency of route determination and the message overhead.

In this protocol, any intermediate node *I* detecting routing failure (due to a broken link) informs the source node *S* by sending a route error packet. Then, *S* initiates a new route discovery (using a request zone) to find a path to the destination *D*. The latency of route redetermination for node *D* can be improved by allowing any intermediate node *I* detecting route error to initiate a route discovery using a request zone based on its own location information for node *D*. Such a local search may result in a smaller request zone because node *I* may be closer to *D* than *S*. Smaller request zone could reduce routing overhead. The time to find the new path to *D* may also be reduced, as a smaller request zone is searched.

7.2.2.9 Query Localization Techniques for On-Demand Routing

The on-demand protocols discover routes via a flooding technique, where the source (or any node seeking the route) floods the entire network with a query packet in search of a route to the destination. Flooding is straightforward to implement. However, as mentioned before, network-wide flooding incurs a considerable overhead and diminishes the performance advantage of on-demand protocols. The primary goal of this technique [30] is to investigate new approaches to reduce the routing overhead by localizing the query flood to a limited region in the network. Similar ideas were explored before. The most prominent among them is the location-aided routing or LAR technique discussed above, which uses the GPS to limit the query flood to a restricted region. However, the approach presented in [30] makes intelligent use of routing histories and does not need location information. On the other hand, it delivers comparable performance advantage.

The proposed query localization protocols are based on the notion of spatial locality: "A mobile node cannot move too far too soon." Thus, prior routing histories can be cached to estimate a small region in the network with high probability of finding the destination node. Only this region needs to be flooded. In this approach, prior route histories are used to limit the query to a region in the neighborhood of the prior routes. Hopwise distance is used to define the neighborhood. The following two heuristics are used in this protocol:

- *Exploiting path locality:* This approach relies on the assumption that after a route to the destination node breaks, the new route cannot be very different than the most recently used route.

- *Exploiting node locality:* The assumption here is that the destination node can be found within a small number of hops from some node on the most recently used route.

It is also possible in rare occasions that the route discovered is not the shortest route. This is possible because there is no guarantee that the shortest route will be contained entirely within the request zone. This will increase the end-to-end delay of the data packets, which will now go via a suboptimal route. However, the performance evaluation shows that the overall savings obtained by the reduction in routing overheads are much more significant and can actually contribute to lower delays.

7.2.3   A Mobile Agent-Based Protocol for Topology Discovery and Routing

7.2.3.1   A Mobile Agent-Based Framework for Topology Discovery

Mobile agents are a novel effective paradigm for distributed applications, and are particularly attractive in a dynamic network environment involving partially connected computing elements. The notion of computation mobility against conventional data mobility governs the underlying philosophy of agencies [31]. Most research examples of the mobile agent paradigm as reported in the current literatures have two general goals: reduction of network traffic and asynchronous interaction. Some authors have suggested that agents can be used to implement network management and to deliver network services [32]. Intensive research on the "Insect-like Systems" has been done over the last few years. The mobile agent systems have been popularly simulated in close resemblance to an ant colony [33]. Of particular interest is a technique for indirect interagent communication, called stigmergy, in which agents populate information cache of nodes, which other agents can use [34]. Stigmergy serves as a robust mechanism for information sharing.

In [35], the issues of geographical-position discovery in ad hoc networks through the implementation of a *mobile multiagent-based framework* have been investigated and a connection management protocol has been proposed. The primary objective of this mobile multiagent framework is to asynchronously collect network topology information (e.g., node velocity, node location) and distribute them into the information cache of all other nodes. In other words, an agent based information exchange and navigation protocol (proactive in nature) has been developed in order to make each node in the network aware of the positions of all other nodes in the network, without consuming large portion of network capacity. GPS support has been used at each node for the extraction of geographical coordinates, velocity, and direction of movement of each node. This location-awareness gives each node an approximate perception of the network topology. Based on this perception, an adaptive routing protocol has been designed ensuring the system to manage an uninterrupted connection in a distributed fashion between any source and destination.

7.2.3.2 Basic Mechanism

Mobile agents or messengers that hop around in the network are a novel solution to the problem of topology discovery. The agents hop from node to node, collect information from these nodes, meet other agents in their journey, interact with both to collect updates of parts of the network that they have not visited or have visited a long time back, and gift these collected data sets to newly visited nodes and agents. A node therefore receives updated information about the network, from the agents visiting them at short regular intervals. Initially when the network commences, all the nodes are just aware of their own neighbors and are in a *don't-know state* regarding the other nods in the system. However with agent navigation beginning, the nodes slowly get information about the other nodes and their neighbors.

For example, let us assume that an agent migrates at every K time tick between nodes. At time t_0, each of the nodes has only information about their immediate neighbors. At time = $t_0 + T$, an agent jumps to a node with the information that it has about its previous host node. Thus this node now has data about a new neighboring node and its neighbors (since the agent has carried this information to it). In the next T time tick, a node gets information regarding two more nodes from another agent. It is to be noted that by controlling T (also known as time to migrate, or TTM), it is possible to control the agent traffic in the network. Moreover, the agent would always migrate from a node to only one of its neighbor after T time-tick. So, the network would never get flooded with propagation of agents [34].

A major aspect underlying the infiltration of topology information into mobile nodes is that the information carried must be recognized with a degree of correctness. Since the agent navigation is asynchronous and there is an obvious time gap between the procurement of information by an agent from one node and its delivery by the same agent to another node, it becomes imperative to introduce a concept of recency of information. For example, let us assume two agents A_1 and A_2 arrive at node n, both of them carrying information about node m which is multihop away from node n. In order to update the topology information at node n about node m, there has to be a mechanism to find out who carries the most recent information about node m: agent A_1 or A_2? To implement that, every node in the network has a counter that is initialized to 0. When an agent leaves a node after completing all its tasks at the node, it increments the counter by one. This counter gives the value of recency. At any point of time, the magnitude of the recency of any node represents the number of times that node was visited by agents since the commencement of the network. This also implies that if two agents have a set of data concerning the same node, say node m, then the agent carrying the higher recency value of node m has more current information about it.

The primary aim of an agent is to collect all topology-related information from its host node in the network and carry them periodically (as updates) to other nodes without flooding the network with topology-update packets. At a periodic interval, an agent propagates its perception of the topology-information to only one of its neighbors, based on a least-visited neighbor-first criterion [34]. At each node, an agent monitors recency of its neighbors to decide which of them has received the least number of agent visits. The neighboring node that has received least number of agent visits so far will be the target node for the agent. Thus, an agent always migrates to a node, which has had the least number of agent visits among the members of the network. Since the agent navigation is a distributed protocol, it attempts to visit all nodes in the network with equal frequency. This in turn facilitates homogeneous topology-awareness.

In Figure 7.4, the text boxes at the bottom displays the time tick, the average topology deviation, the number of agents in the system, the number of nodes, the transmission range and the mobility of the system. In the inset, the local topology perception of the node 9 is shown. This window has been invoked by clicking on node 9 during the execution of the simulator. The nodes shown in black in the inset represents the positions of the nodes as perceived by node 9 and the nodes in white depict the actual position of the nodes in the network topology.

7.2.3.3   Uninterrupted Connection Management

This topology-awareness discussed above leads to a scenario where conventional route discovery is no longer necessary. More explicitly, the nodes can now determine the most stable route locally and initiate the sending of data packets through it. The way to calculate the stability of a path is indicated [24]. After a point of time, if the source node finds that the chosen route has attained a low stability (indicating that it would soon cease to exist), the node computes a new, more stable route from the local information cache and redirects data packets through the later. This adaptive route selection facilitates continuous communication through multiple paths in the temporal domain. Thus, as long as two nodes remain connected, they will always be able to get at least one route through which communication can continue. In the case of multiroute availability, the best route can always be selected. Quite perceivably, the adaptive selection of best routes guarantees an uninterrupted communication session between two nodes thus ensuring multimedia data transfer to occur.

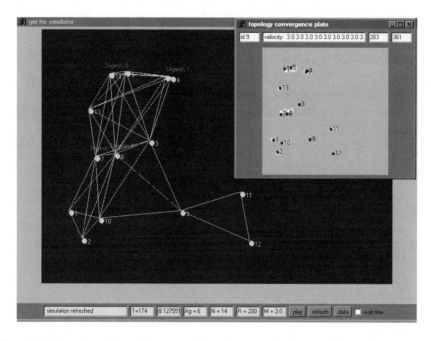

Figure 7.4   Snapshot of the ad hoc network running on a simulator with a set of 14 nodes.

From the above discussion we see that a node initiates and even redirects message communication through appropriate routes without consuming any network bandwidth. A route error can occur when either the perceived route does not exist at all or when the perceived stability is substantially higher in comparison to the actual route stability. The later case indicates that the node is misled to believe that it would remain connected to its destination (through a particular route) for a specific span of time T whereas this path would actually cease earlier. Now both these situations would be the result of inaccurate topology information. In other words, if the node's topology information is significantly incongruous with the actual topology of the ad hoc architecture, the deviation of the perceived path and the perceived stability would be proportionally deviant from the actual path and stability, thus increasing the route error frequency.

This does not pose to be a cause of serious concern in this proposal. First of all, even if a route error occurs, it does not necessitate a route-rediscovery process. The source node merely selects another route from its route cache. Secondly, the route error frequency bears a direct relation with the average topology deviation (average deviation between actual topology and perceived topology at each node) and thus indirectly with the performance of the agent-based framework. In a low-mobility system or a low-TTM system, the average topology deviation is significantly lower and thus the difference between the perceived topology and the actual is less. If the source node takes care of this fact during adaptive route selection, the chance of route errors would be reduced.

7.2.4 Power-Aware Routing Protocols in Ad Hoc Networks

7.2.4.1 Transmission Power Control in Ad Hoc Networks

Past work on power control has primarily dealt with cellular networks for managing cochannel interference at receivers and increasing network capacity through frequency reuse [36]. Essential to power control in this context is online link quality monitoring for adaptation to changes induced by mobility and channel impairments. Several power control algorithms have been presented with the following objectives:

- Increased power savings at mobile nodes;

- Better maintenance of stable QoS on wireless links;

- More efficient handling of mobility (structural stretching and hand-offs);

- Reduced interference on other links sharing the channel to enhance network capacity.

The use of a power-control paradigm in ad hoc networks is relatively new. The problem of minimum energy routing or power-aware routing in ad hoc networks has been addressed in [37, 38]. One of the approaches has been to minimize the total consumed energy in reaching the destination, which minimizes the energy consumed per unit flow or packet [37]. This approach may drain out the batteries of certain paths, which may disable further information delivery even though there are many nodes with plenty of energy. Another approach is to maximize the system lifetime given the sets of origin and destination nodes and the information generation rate at the origin nodes. In yet another approach [39], a new multiaccess protocol has been proposed. Radios that are not actively transmitting or receiving a packet switch themselves off without affecting the throughput. All these approaches have been shown to achieve their objective of energy conservation; however, they do not concern themselves with the mechanism of transmit power control to optimize network performance.

Recently, two power-controlled MAC protocols (PCMAs) have been proposed [40, 41] in ad hoc networks. In [40], a PCMA is proposed where a sender will use an appropriate power level to transmit its packets so as to increases the channel utilization and save battery power. They have shown that the channel utilization can be significantly increased because the severity of signal overlapping is reduced. In [41], a PCMA within the collision avoidance framework has been proposed. They have demonstrated that PCMA can allow more number of simultaneous communication than 802.11 by adapting the transmission ranges to the minimum value required to satisfy successful reception at the intended destination. However, the main focus of both the works is to demonstrate the channel utilization using transmit power control. How to maintain a topology and how to incorporate a reactive routing scheme using power control has not been addressed. For example, connectivity in an ad hoc network needs to be maintained with suitable power control mechanism, even if there is no communication. And, a reactive routing protocol has to operate on that controlled topology. To address this problem, several researchers have investigated the issue of topology control in ad hoc networks, as discussed below.

7.2.4.2　Topology Control in Ad Hoc Networks by Adjusting Transmission Power

The topology of an ad hoc wireless network is the set of communication links between node pairs and is primarily dependent on the transmission range of individual nodes. Normally, it is assumed that the transmission range is fixed and is a given parameter. However, it has been observed that a low transmission range will not guarantee proper connectivity among mobile hosts to ensure effective communication. On the other hand, if the transmission range is high, it will ensure connectivity but will decrease channel utilization [40] and increase collision and congestion of control packets [24], which will increase the end-to-end delay significantly. Hence, for a fixed transmission range, if the node density is low, the network would be partitioned into disjoint components and proper network connectivity cannot be ensured. On the other hand, if the node density is high, poor channel utilization and congestion and collision due to control packets will degrade the network performance.

Adaptive adjustment of transmission power of each individual node can control the topology of the network [42, 43]. Here, the objective of each node in the network is to keep its number of neighbors bounded. Usually, during route-discovery phase, a source node broadcast a route-request packet that propagates from node to node and mainly responsible in generating the large number of control packets in the system. If G_{avg} is the average number of neighbors of a node and if the maximum number of hops is limited to 4, then the approximate number of route-request control packets generated will be G^4_{avg} communication event. Therefore, it is obvious that with increase in G_{avg}, the number of route-request control packets increases drastically. For a fixed transmission range, the number of control packets generated increases drastically beyond a certain node density, as it increases G_{avg}. However, if the transmission range is adjusted to keep the average number of neighbors bounded, the number of route-request control packets generated would be uniform. Moreover, control of transmission range will improve channel utilization by allowing more number of simultaneous communications. At the same time, this strategy will also help to increase the probability of a connected network to a large extent.

In a recent work on topology control, Ramanathan and Rosales-Hain [43] have addressed the issue of topology control in ad hoc wireless network using transmit power adjustment. For mobile ad hoc networks, they have presented two distributed heuristics that adaptively adjust a node's transmit powers in response to topological changes. Their results show that the performance of multihop wireless networks in practice can be substantially increased with topology control. In our paper, we will discuss only on the first heuristic, namely local information no topology (LINT). The second one, local information link-state topology (LILT), depends on the global topology information that needs to be available with some proactive routing protocols such LSPs. However, many researchers feel that the proactive protocols, where each node tries to capture the global topology information, are not suitable in highly dynamic environment of ad hoc networks. Moreover, in LILT, an excessive number of topology control-induced updates will be generated that will propagate globally and decrease the throughput of the network. Hence, we will not take into account LILT in our paper and discuss only LINT.

LINT attempts to keep the degree (number of neighbors) of each node bounded. If the node degree is less than certain threshold, the node will increase its transmission range. If the node degree is more than certain threshold, it will decrease its transmission range. Thus, the basic mechanism of LINT depends on the calculation of a node degree. It is assumed that the number of neighbors (degree) of any node will be known from its neighborhood table (built by some routing mechanism). It has also been assumed that some neighborhood discovery protocol will return only bidirectional links. However, there is a significant shortcoming in these assumptions. In all reactive routing in ad hoc network, a node n will be aware of its neighbor only through a link-level beacon exchange from its neighbors. In other words, each node periodically broadcast a beacon to notify its existence. Thus, the number of beacon a node n will receive will determine the neighborhood counts of n, provided those nodes are also reachable from n (to ensure bidirectionality). At the same time, the number of beacons a node n will receive does not depend on the transmission range of n but on the transmission range of other nodes. So, even if we increase the transmission range of n, it does not imply that it would increase the number of neighbors of n. As a consequence, a node in the system may increase its transmission range indefinitely without increasing its number of bidirectional neighbors, simply because other nodes may not have any incentive to increase their transmission range. So, degree of a node n can only be calculated and manipulated, if there is a kind of handshaking between a node n and its intended neighbors to establish a neighborhood agreement. As a result, both n and its prospective neighbors have to manipulate their transmission ranges in order to maintain a desired number of neighborhood relationships [42].

Another limitation of LINT is in its assumption of uniform node distribution. Based on this assumption, the targeted value of transmit power for each node to attain a desired degree is calculated. However, uniform distribution is not a realistic assumption. Moreover, no adaptive mechanism has been incorporated to correct or rectify this assumption. We have incorporated an adaptive mechanism to adjust transmission range until the desired degree (in bidirectional sense) is actually achieved.

7.2.4.3 Selecting Optimal Transmission Range

The selection of optimal transmission range to maximize throughput and ensure connectivity is studied in [44, 45]. Under certain assumptions, they have derived an optimal node-degree, called magic number. Assuming uniform distribution, if the average node density is λ and A_R is the area of coverage of node i for transmission range R ($= \Pi R^2$), then the optimal value of $\lambda . A_R$ which maximizes throughput has been referred to as magic number. It has been shown that the value of $\lambda . AR$ from six to eight will have a high probability of leading to a connected network as well as reducing congestion and collision in the network. Philips et al. [46] further conjectured that given a very large area, no fixed optimal number as magic number can exists. To ensure network connectivity, it has been shown that the expected number of nearest neighbor (node degree) must grow logarithmically with the area of the network. They have shown that node degree should be between 2.195 and 10.526 to have a connected network with nonzero probability. Gupta et al. [47] addressed asymptotic connectivity of ad hoc network under infinite node environment. Kirousis et al. [48] studied the problem of assigning transmission ranges to the nodes of a multi hop network so as to minimize the total power consumption under the constraint of ensuring a connected network. Sanchez et al. [49] and Elizabeth et al. [50] reported simulation results on critical range and impact of transmission range on throughput. All the analytical studies on connectivity assume uniform distribution. The problem becomes analytically intractable in case of nonhomogeneous distribution of nodes. Moreover, if there is a large number of nodes in a large area, maintaining connectivity with very high probability can lead to a large transmission range [51].

For a fixed number of nodes uniformly distributed over an operating area, an optimal transmission range can be worked out. But this would give a theoretical solution only. In practice, if we consider a dynamic ad hoc network environment, the number of nodes as well as the concentration of node in different area of the operating zone varies. Therefore, a single, optimal transmission range, even if we find it, will not be a solution in this context. A protocol based on variable transmission range would be highly effective in such a dynamic environment. It has been observed that [42] when nodes are randomly distributed in a finite plane with nonhomogeneous node distribution and the node degree is maintained between 4 and 8 (with an average value of 6, the magic number) by adjusting the transmission range of individual node, it can ensure a connected network with a probability of 96%.

7.2.4.4   A Routing Protocol with Adaptive Transmission Range Control

A routing protocol proposed with adaptive transmission range control [42] is an on-demand reactive protocol. The protocol has four components:

1. *Transmission Range Control Protocol* (TRCP), which will help each node to adjust periodically its transmission range in order to maintain the number of its registered neighbors within four to eight;

2. *Power-Sensitive MAC Protocol* (PMAC), which controls medium access in an environment where transmission range of each user terminal is variable;

3. *Path Finding Protocol* (PFP), which is a demand-driven, source-initiated routing protocol to find a set of paths between any source-destination pair;

4. *Path Evaluation and Data Communication Protocol* (PEDCP), which helps the source to evaluate the best path among the set of paths and to start the data communication along that path. Additionally, it helps TRCP to protect the links in the path during data communication.

The proposed protocol is based on a neighborhood agreement and denial scheme through periodic beacon exchange among the neighboring nodes only. The basic assumption is that, even if a node is within the transmission range of another node, it will not be considered as its neighbor unless both of them agree to be the neighbor of each other. This ensures bidirectionality of neighborhood relationship. On the other hand, a node can increase its transmission range to include someone as its neighbor. A node will vary its transmission range so that the average number of its registered neighbor is around six. It has been pointed out earlier [44, 45] that if the average number of neighbors were six (the magic number), it would optimize the throughput by reducing congestion and collision as well as ensuring connectivity of the network.

During routing of data packets, a node-pair will try to maintain their neighborhood relationship by adjusting the transmission range, that is, protect the link between them, if that link is involved in a communication process at that instant of time. A source would evaluate a set of paths after route discovery and find out the best path for data communication. The best path is a path where the path protection mechanism during data communication will consume the least battery power during the adjustments of their transmission range. This is in tune with the concept of active link protection with power control in cellular network [52]. In those schemes, the transmitter power is adjusted in order to maintain the link quality and preserve the network topology under structural stretching (elastic network). One of the problem is that, some powers may increase excessively and links become too expensive to maintain. That is why a path evaluation mechanism is used to find out the best path in this context.

7.2.5   Other Routing Protocols

Zone Routing Protocol (ZRP) [53] is a zone- or cluster-based hybrid routing protocol suitable for a wide variety of mobile ad hoc networks, especially those with large network spans and diverse mobility patterns. Each node proactively maintains routes within a local region (referred to as the routing zone). Knowledge of the routing zone topology is leveraged by the ZRP to improve the efficiency of a reactive route query and reply mechanism. The proactive maintenance of routing zones also helps improve the quality of discovered routes, by making them more robust to changes in network topology. The ZRP can be configured for a particular network by proper selection of a single parameter, the routing zone radius. However, ZRP should be viewed more as a routing framework rather than an independent protocol, as potentially any proactive protocol can be employed for intrazone (or, intracluster) routing and any reactive protocol can be employed for interzone (or, intercluster) routing.

In ZRP, each node has its own routing zone, which includes the nodes whose distance (hops) is at most some predefined number. Each node is required to know the topology of the network within its routing zone only, and route updates are propagated only within the routing zone. A proactive protocol such as DSDV is used within the routing zone to learn about its topology. To discover a route to an out-of-zone nodes, a reactive protocol such as DSR is used. Note that ZRP exhibits hybrid behavior of proactive and reactive through the use of the zone radius. For large zone radius, ZRP is more proactive, and for small zone radius, ZRP is more reactive. The route discovery requires a relatively small number of query messages as these messages are routed only to peripheral nodes, omitting all the nodes within the routing zones. This method is called bordercasting.

The advantage of ZRP is that it significantly reduces the communication overhead as compared to the pure proactive protocols since in ZRP each node needs to know the topology of its zone only. In addition, ZRP discovers routes faster than the pure reactive protocols since in ZRP only the peripheral nodes are queried in the route discovery process.

Basic and Hierarchical Spine Routing   is a distributed algorithm used to find an approximation to the minimum connected dominating set (MCDS) of an ad hoc network [54]. Topological information is then gathered from non-MCDS nodes and propagated to all MCDS nodes. Each MCDS node runs an all-pair shortest path algorithm. A non-MCDS node can then request the route information to a destination from an adjacent MCDS node. Since MCDS nodes propagate, store and run shortest path algorithm on the topological information, their communication, storage and computation burdens are heavy. Moreover, mobility management is complicated by the existence of MCDS nodes.

Hierarchical spine routing (HSR)   is a two-level routing architecture used to solve the scalability problem in basic spine routing (BSR). The network is divided into clusters. Each cluster head stores the topology of a cluster graph, the cluster membership table (it consists of cluster ID for each node) and the list of local boundary nodes (it shows clusters reachable for each boundary node). Within each cluster, BSR is used. Between clusters, link-state routing is run on the cluster graph topology. Using distinct clustering, the impact on topological change is localized. Moreover, the amount of time for spine construction and maintenance is reduced since clusters essentially run in parallel. The communication and the storage overhead in some nodes can be greatly reduced. However, cluster heads, boundary nodes, and spine nodes take on heavy responsibilities. Therefore, the communication, storage and computation burdens are heavy in those nodes. Mobility management becomes complicated and the reliability of the network decreases due to single point of failure.

Optimized Link-State Routing (OLSR) Protocol [55]   is an optimization of the pure link-state algorithm tailored to the requirements of a mobile WLAN. The key concept used in the protocol is that of multipoint relays (MPRs). MPRs are the selected nodes, which forward the broadcast packets during the flooding process. This technique substantially reduces the packet overhead as compared to pure flooding mechanism where every node retransmits the packet when it receives the first copy of the packet. The protocol is particularly suitable for the large dense networks as the technique of multipoint relays works well in this context.

Relative Distance Microdiscovery Ad Hoc Routing (RDMAR) Protocol [56]   is highly adaptive, bandwidth-efficient and scaleable. A key concept in its design is that protocol reaction to link failures is typically localized to a very small region of the network near the change. This desirable behavior is achieved through the use of a novel mechanism for route discovery, called relative distance microdiscovery (RDM). The concept behind RDM is that a query flood can be localized by knowing the relative distance (RD) between two terminals. Based on the RD, the query flood is then localized to a limited region of the network centered at the source node of the route discovery and with maximum propagation radius that equals to the estimated relative distance. This ability to localize query flooding into a limited area of the network serves to increase scalability and minimize routing overhead and overall network congestion.

7.3 Multicast Routing Protocols in Ad Hoc Networks

Multicast communication in the context of ad hoc networks is a very useful and efficient means of supporting group-oriented applications, where the need for one-to-many data dissemination is quite frequent in critical situations such as disaster recovery or battlefield scenarios. Instead of sending data via multiple unicasts, multicasting reduces the communication costs by minimizing the link bandwidth consumption and delivery delay [57].

Research in the area of routing in ad hoc wireless network has mostly concentrated on designing effective routing schemes for unicast communication. Those routing algorithms were not designed with multicast extensions in mind. Therefore, they do not naturally support multicast routing solutions. Since fixed network multicasting is based on state in routers (either hard or soft), it is fundamentally unsuitable for ad hoc network where topology is changing frequently due to unconstrained mobility. It has been shown that the performance of both hard- and soft-state multicast tree maintenance mechanisms degrade rapidly with increased mobility. Multicast protocols used in static networks, such as *Distance Vector Multicast Routing Protocol* (DVMRP) [58, 59], *multicast open shorted path first* (MOSPF) [60], *core-based trees* (CBT) [61], *and protocol independent multicast* (PIM) [62], do not perform well in ad hoc networks because multicast tree structures are fragile and must be readjusted as connectivity changes. Furthermore, multicast trees usually require a global routing substructure such as link-state or distance vector. The frequent exchange of routing vectors or link-state tables, triggered by continuous topology changes, yields excessive channel and processing overhead. Thus, traditional multicast approaches that rely on maintaining and exchanging multicast-related state information are not suitable in highly dynamic ad hoc network with frequent and unpredictable changing topology.

To construct a multicast delivery path, multicast routing schemes need to provide a way for the delivery path to adapt to changes in network topology as well as to changes in group membership. They need to do this in a timely fashion, not wasting bandwidth by sending unwanted data to resigned members or failing to forward multicast traffic to new members. State retention overhead for routing needs to be minimized at each host.

To do multicasting, some way is needed to define multicast groups. In conventional multicasting algorithms, a multicast group is considered as a collection of hosts, which register to that group. It means that, if a host wants to receive a multicast message, it has to join a particular group first. When hosts want to send a message to such a group, they simply multicast the message to the address of that group. All the group members then receive the message. An easily reconfigurable multicast tree topology is required since it can be dynamically changed by movement of group members. However, unfortunately, channel overhead caused by tree reconfiguration updates tends to increase very rapidly with mobility [63]. Sometimes, multicast flooding may be a better solution in ad hoc networks.

Currently proposed ad hoc multicast routing schemes lie on a spectrum that spans from pure Internet multicast routing based schemes to a pure flooding scheme. Internet multicast routing schemes, as it is currently, generally require the routing nodes to maintain fairly large amount of state information for routing and to use processing power of hosts rather liberally. Feasibility of supporting continuous unlimited mobility is also a question with Internet routing schemes. Only flooding control packets may support unlimited continuous mobility. Flooding will also reduce the amount of state information kept at mobile hosts, and will provide reliable and timely delivery.

The *Forwarding Group Multicast Protocol* (FGMP) [64] proposes a scheme that is a hybrid between flooding and source based tree multicast. The proposed multicast protocol scheme keeps track not of links but of groups of nodes, which participate in multicast packets forwarding. To each multicast group *G* is associated a forwarding group *FG*. Any node in *FG* is in charge of forwarding (broadcast) multicast packets of *G*. The nodes to be included in *FG* are elected according to members' requests. Instead of data packets, small membership advertisement packets are used to reduce overhead caused by broadcasting. However, in order to advertise the membership, each receiver periodically and globally flood its member information.

The *Ad Hoc Multicast Routing Protocol (AMRoute)* [65] creates a per-group multicast distribution tree using unicast tunnels connecting group members. The protocol has two main components: mesh creation and tree creation. Certain nodes are designated as logical core nodes that initiate mesh and tree creation; however, the core can migrate dynamically according to group membership and network connectivity. Logical cores are responsible for initiating and managing the signaling component of AMRoute, such as detection of group members and tree set up. Bidirectional tunnels are created between pairs of group members that are close together, thus forming a mesh. Using a subset of available mesh links, the protocol periodically creates a multicast distribution tree. AmRoute assumes the existence of an underlying unicast routing protocol and its performance is influenced by the characteristics of the unicast routing protocol being used. The AMRoute simulation runs on top of TORA [21] as underlying unicast protocol. The network dynamicity was emulated by keeping node location fixed and breaking/connecting links between neighboring nodes. Thus, the effect of actual node mobility on the performance is difficult to interpret. Moreover, the signaling generated by underlying unicast protocol (TORA in this case) is not considered in the measurements.

The *Core-Assisted Mesh Protocol* (CAMP) [66] generalizes the notion of CBTs introduced for Internet multicasting into multicast meshes that have much richer connectivity than trees. A shared multicast mesh is defined for each multicast group. The advantage of using such meshes is to maintain the connectivity even while the network routers move frequently. CAMP consists of the maintenance of multicast meshes and loop-free packet forwarding over such meshes. Multicast packets for a group are forwarded along the shortest path from sources to receivers defined within the group's mesh. CAMP rebuilds meshes at least as fast as CBT and PIM can rebuild trees. However, the effect of mobility on the performance has not been clearly evaluated. The topology under experimentation has 30 routers with high connectivity (an average of 6 neighbors each) and at the most 15 routers out of 30 are assumed to be mobile.

The *ad hoc multicast routing protocol utilizing increasing ID numbers,* (AMRIS) [67] assigns an identifier to each node in a multicast session. A premulticast session delivery tree rooted at a special node (by necessity a sender) in the session joins all the group members. The tree structure is maintained by assigning identifiers in increasing order from the tree root outward to the other group members. All nodes are required to process the tree set up and maintenance messages that are transmitted by the root periodically. It has been assumed that most multicast applications are long-lived; therefore rapid route reconstruction is of greater importance compared to rapid route discovery. The performance of the proposed scheme has yet to be evaluated.

The *On-Demand Multicast Routing Protocol,* (ODMRP) [68] also uses a mesh-based approach for data delivery and uses a forwarding group concept. It requires sources rather than destinations to initiate the mesh building by periodic flooding of control packets. It applies on-demand procedures to dynamically build routes and maintain multicast group membership. A soft-state approach is taken to maintain multicast group member. In networks where GPS is available, ODMRP can be made adaptive to node movements by utilizing mobility prediction. By using location and mobility information supported by GPS, route expiration time can be estimated and receivers can select the path that will remain valid for the longest time. With the mobility prediction method, sources can reconstruct routes in anticipation of route breaks. This way, the protocol becomes more resilient to mobility. The price is, of course, the cost and additional weight of GPS.

A *location-based multicast* scheme has been proposed in [69] to decrease delivery overhead of multicast packets, as compared to multicast flooding. A location-based multicast group has been defined as the set of nodes residing in a geographical multicast region. Thus, if a host resides within a specific area at a given time, it will be automatically a member of the corresponding multicast group at that time. All the hosts in the multicast region should receive the multicast packet. Such a multicast group may be used for sending a message that is likely to be of interest to everyone at a given location (or in a specified area). The scheme attempts to reduce the forwarding space for multicast packets. Limiting the forwarding space results in fewer multicast messages, while maintaining the accuracy of data delivery comparable with multicast flooding.

Stability-based multicast routing [24] is based on the stability-based unicast routing scheme described above. The mechanism comprises four sequential steps. First, source initiates a route discovery to get all the paths to individual destinations; next, it selects the stable routes from them and constructs a subgraph connecting source and destinations; next, source extracts multicast tree(s) from this subgraph; finally, source communicates data to destinations using the multicast tree(s). The proposed multicast routing mechanism depends only on local state information at source for constructing a multicast tree dynamically and is demand-driven in the sense that whenever a source needs to communicate with a set of destinations belonging to a multicast group, it discovers the routes to the individual destinations and creates a multicast tree dynamically at source for that given group. Stability-based multicast routing scheme proposed here will ensure that the lifespan of the multicast tree so formed will be sufficient to complete the required volume of data transfer at that instant of time.

7.4 Performance Comparisons of Unicast and Multicast Routing Protocols

The comparative analysis of major unicast routing protocols has been studied extensively through simulation and performance evaluation in [15, 705073]. The performance comparison study of ad hoc wireless multicast protocols has been presented in [63]. In this section, we will discuss the summary of their findings.

7.4.1 Performance Comparisons of Major Unicast Routing Protocols

The performance of DSR and AODV, two prominent on-demand routing protocols for ad hoc networks has been evaluated in [15]. A detailed simulation model has been developed to demonstrate the performance characteristics of the two protocols. The general observation from the simulation is that for application-oriented metrics such as delay and throughput, DSR outperforms

AODV in less stressful situations (i.e., a smaller number of nodes and lower load or mobility or both). AODV, however, outperforms DSR in more stressful situations, with widening performance gaps with increasing stress (e.g., more load, higher mobility). DSR, however, consistently generates less routing load than AODV. The poor delay and throughput performances of DSR are mainly attributed to aggressive use of caching, and lack of any mechanism to expire stale routes or to determine the freshness of routes when multiple choices are available. The mechanisms to expire routes and/or determine freshness of routes, will benefit DSR's performance significantly. Since AODV keeps track of actively used routes, multiple actively used destinations also can be searched using a single route discovery flood to control routing load. In general, it was observed that both protocols could benefit (1) from using congestion-related metrics (such as queue lengths) to evaluate routes instead of emphasizing the hopwise shortest routes, and (2) by removing "aged" packets from the network. The aged packets are typically not important for the upper layer protocol, because they will probably be retransmitted. These stale packets do contribute unnecessarily to the load in the routing layer.

In [71], a comprehensive performance evaluation of major proactive and reactive routing protocols has been presented. Steady state performance in terms of fraction of packets delivered, delay and routing load have been considered as the performance metrics. Even with a packet-level simulation model the essential aspects of the routing protocols are exposed. The key observations are as follows. The proactive, shortest path protocols provide excellent performance in terms of end-to-end delays and packet delivery fraction at the cost of higher routing load. The on-demand protocols suffer from suboptimal routes as well as worse packet delivery fraction because of more dropped packets. However, they are significantly more efficient in terms of the routing load. The multipath protocol, TORA, did not perform well in spite of maintaining multiple redundant paths. The overhead of finding and maintaining multiple paths and the protocol's sensitivity to the loss of routing packets seem to outweigh the benefits of multiple paths. Also, the end-to-end delay performance is poor because of the loss of distance information. The routing load differentials between all routing protocols reduce with large number of peer-to-peer conversations in the network. However, the other performance differentials are not affected conclusively.

In this work [70], the results of a detailed packet-level simulation comparing four recent multihop wireless ad hoc network routing protocols has been presented. These protocols, DSDV, TORA, DSR, and AODV, cover a range of design choices, including periodic advertisements versus on-demand route discovery, use of feedback from the MAC layer to indicate a failure to forward a packet to the next hop, and hop-by-hop routing versus source routing. They have simulated each protocol in ad hoc networks of 50 mobile nodes moving about and communicating with each other, and presented the results for a range of node mobility rates and movement speeds.

Each of the protocols studied performs well in some cases yet has certain drawbacks in others. DSDV performs quite predictably, delivering virtually all data packets when node mobility rate and movement speed are low, and failing to converge as node mobility increases. TORA, although the worst performer in the experiments conducted in terms of routing packet overhead, still delivered over 90% of the packets in scenarios with 10 or 20 sources. At 30 sources, the network was unable to handle all of the traffic generated by the routing protocol and a significant fraction of data packets were dropped. The performance of DSR was very good at all mobility rates and movement speeds, although its use of source routing increases the number of routing overhead bytes required by the protocol. Finally, AODV performs almost as well as DSR at all mobility rates and movement speeds and accomplishes its goal of eliminating source routing overhead, but it still requires the transmission of many routing overhead packets and at high rates of node mobility is actually more expensive than DSR.

7.4.2   Performance Comparisons of Major Multicast Routing Protocols

A performance evaluation and comparative analysis of five multicast protocols for ad hoc networks has been presented in [63]. The channel, radio, IEEE 802.11 MAC protocol, and multicast protocols (AMRoute, ODMRP, AMRIS, CAMP, and flooding) have been carefully implemented The simulator has modeled a realistic wireless environment and results have been obtained for a broad range of parameters including mobility, number of senders, multicast group size, and traffic load. A general conclusion is that, in a mobile scenario, mesh-based protocols outperformed tree-based protocols. The availability of alternate routes provided robustness to mobility. AMRoute performed well under no mobility, but it suffered from loops and inefficient trees even for low mobility. AMRIS was effective in a light traffic environment with no mobility, but its performance was susceptible to traffic load and mobility. CAMP showed better performance when compared to tree protocols, but with mobility, excessive control overhead caused congestion and collisions that resulted in performance degradation. ODMRP was very effective and efficient in most of the simulation scenarios. However, the protocol showed a trend of rapidly increasing overhead as the number of senders increased.

7.5   Discussion

Several researchers feel that on-demand reactive routing schemes that do not use periodic message of any kind would be more suitable in the context of ad hoc networks. However, it has been observed that these protocols perform well under light traffic and low mobility, but performance degrades significantly under high mobility and high traffic load. As mobility increases, the precomputed route may break down, requiring multiple route discoveries on the way to destination. Route caching becomes ineffective in high mobility. Stability-based routing schemes that tend to evaluate the lifespan of a path reduce this problem, but cannot eliminate it. Moreover, since flooding is used for query dissemination and route maintenance, on-demand routing tends to become inefficient when the frequency of communication requirement is high.

On the other hand, several researchers have proposed proactive routing schemes based on classical DVR or LSR. The DVR approach is simple and communication overhead is less than that of LSR, but it suffers from slow convergence and the tendency to create routing loops. In LSR, global network topology information is maintained in all routers by the periodic flooding of link-state updates by each node. Any link change triggers an immediate update. As a result, the time required for a router to converge to the new topology is much less than in the distance vector approach. Due to global topology knowledge, preventing routing loop is also easier. Moreover, the link-state information to evaluate the quality of the entire path (e.g., bandwidth, delay, lifespan, affinity) can be easily obtained from the global topology information maintained at each node. Unfortunately, as these schemes rely on flooding of routing updates, excessive control overhead may be generated, especially in a highly mobile environment. Thus, in the context of ad hoc networks, researchers have focused on restricting the propagation of routing updates, thereby reducing the control overheads. For example, OLSR protocol is an optimization of the pure link-state algorithm tailored to the requirements of a mobile WLAN. The key concept used in the protocol is that of MPRs. MPRs are the selected nodes, which forward the broadcast packets during the flooding process. This technique substantially reduces the packet overhead as compared to pure flooding mechanism where every node retransmits the packet when it receives the first copy of the packet. FSR introduces the notion of multilevel fisheye scope to reduce routing update overhead in large networks. Nodes exchange link-state entries with their neighbors with a frequency that depends on distance to destination. From link-state entries, nodes construct the topology map of the entire network and compute optimal routes.

Designing a proper routing scheme to suit different application contexts of ad hoc networks for efficient message communication along with power conservation is still a challenging task. With the advent of radios with more sophisticated controls on transmission parameters, adaptive transmission range is going to be a viable commercial solution in near future. An energy-conserving routing protocol with adaptive transmission range control may be a good candidate for future ad hoc networks.

References

[1]   Perlman R., *Interconnections: Bridges and Routers,* Addison-Wesley, 1992, pp. 149–152, 205–233.

[2]   Jubin, J., and J. Tornow, "The DARPA Packet Radio Network Protocols," *Proc. of the IEEE* 75(1), Jan. 21–32, 1987.

[3]   McQuillan, John M., Ira Richer, and Eric C. Rosen, "The New Routing Algorithm for the ARPANET," *IEEE Transactions on Communications,* Vol. 28, No. 5, May 1980, pp. 711–719.

[4]   Moy, J., "Multicast Routing Extensions for OSPF," *Communications of the ACM,* Vol. 37, No. 8, Aug. 1994, pp. 61–66, 114.

[5]   Corson, S., J. Macker, and S. Batsell, "Architectural Considerations for Mobile Mesh Networking," *Internet Draft RFC Version 2,* May 1996.

[6]   Johnson, D., "Routing in Ad Hoc Networks of Mobile Hosts," *Proc. IEEE Workshop on Mobile Comp. Systems and Appls.,* Dec. 1994.

[7]   Pei, Guangyu, Mario Gerla, and Tsu-Wei Chen, "Fisheye State Routing: A Routing Scheme for Ad Hoc Wireless Networks," *Proc. of the IEEE International Conference on Communication,* New Orleans, LA, June 2000.

[8]   Royer, E. M., and C. K. Toh, "A Review of Current Routing Protocols for Ad Hoc Mobile Wireless Networks," *IEEE Personal Communications,* Apr. 1999.

[9]   Perkins, E., and P. Bhagwat, "Highly Dynamic Destination-Sequenced Distance-Vector Routing (DSDV) for Mobile Computers," *ACM Comput. Commun. Rev.,* Vol. 24, No. 4, (ACM SIGCOMM'94) Oct. 1994, pp. 234–244.

[10] Murthy, S., and J. J. Garcia-Luna-Aceves, "An Efficient Routing Protocol for Wireless Networks," *ACM Mobile Networks and Applications J.,* Vol. 1, No. 2, 1996, pp. 183–197.

[11] Krishna, P., et al., "A Cluster-Based Approach for Routing in Ad Hoc Networks," *Proc. of the 2nd USENIX Symp. on Mobile & Location-Independent Computing,* Apr. 1995.

[12] Chiang, C. C., and M. Gerla, "Routing in Clustered Multihop Mobile Wireless Networks," *Proc. of 11th International Conference on Information Networking,* Taiwan 1997.

[13] Kleinrock, L., and K. Stevens, "Fisheye: A Lenslike Computer Display Transformation," technical report, UCLA, Computer Science Department, 1971.

[14] Johnson, B., and D. A. Maltz, "Dynamic Source Routing in Ad Hoc Wireless Networks," T. Imielinski and H. Korth, (eds.) *in Mobile Computing,* Boston: Kluwer, 1996.

[15] Das, S R, Charles Perkins and Elizabeth Royer, "Performance Comparison of Two On-Demand Routing Protocols for Ad Hoc Networks," *Proc. of the IEEE INFOCOM 2000,* Tel Aviv, Israel, March 26–30, 2000.

[16] Park, V. D., and M. S. Corson, "A Highly Adaptive Distributed Routing Algorithm for Mobile Wireless Networks," *IEEE INFOCOM'97,* Kobe, Japan, Apr. 1997.

[17] Perkins, E., "Ad Hoc On-Demand Distance Vector (AODV) Routing," Internet Draft, Nov. 1997.

[18] Perkins, C. E., and E. M. Royer, "Ad Hoc On-Demand Distance Vector Routing", *Proc. of the 2nd IEEE Workshop on Mobile Comp. Sys. And Apps.*, Feb. 1999, pp. 90–100.

[19] Perkins, Charles, Elizabeth Royer, and Samir Das. "Ad Hoc On-Demand Distance Vector (AODV) Routing," http:// www.ietf.org/ internet-drafts/ draft-ietf-manet-aodv-03.txt, June 1999, IETF Internet Draft (work in progress).

[20] Gafni, E., and D. Bertsekas, "Distributed Algorithms for Generating Loop-Free Routes in Networks with Frequently Changing Topology," *IEEE Trans. Commun.*, Jan. 1981.

[21] Corson, M. S., and A. Ephremides, "A Distributed Routing Algorithm for Mobile Wireless Networks," *ACM/Baltzer Journal on Wireless Networks*, Vol. 1, No. 1, Feb. 1995, pp. 61–82.

[22] Toh, C. K., "A Novel Distributed Routing Protocol to Support Ad Hoc Mobile Computing," *IEEE International Phoenix Conf. on Computer & Communications* (IPCCC'96), 1996.

[23] Dube, R., et al., "Signal Stability Based Adaptive Routing for Ad Hoc Mobile Networks," Technical Report CS-TR-3646, UMIACS-TR-96-34, Institute for Advanced Computer Studies, Department of Computer Science, University of Maryland, College Park, MD, Aug. 1996.

[24] Paul, Krishna, et al., "A Stability-Based Distributed Routing Mechanism to Support Unicast and Multicast Routing in Ad Hoc Wireless Network," *Computer Communications (Elsevier Science)*, Vol. 24, Dec. 2001, pp. 1828–1845.

[25] Paul, Krishna, Romit Roychoudhuri, and S. Bandyopadhyay, "Survivability Analysis of Ad Hoc Wireless Network Architecture," *Proc. of the Second International Workshop on Mobile and Wireless Communication Networks*, Paris, France, May 2000. Published in Lecture Notes in Computer Science LNCS 1818, Springer Verlag.

[26] Das, Sajal K., et al., "Improving Quality-of-Service in Ad Hoc Wireless Networks with Adaptive Multipath Routing," *Accepted in GLOBECOM 2000*, San Francisco, CA, Nov. 27–Dec 1, 2000.

[27] Fisher, M. L., "An Applications Oriented Guide to Langrangean Relaxation," *Interfaces*, Vol. 15, 1985, pp. 10–21.

[28] Ko, Y. B., and N. H. Vaidya, "Location-Aided Routing in Mobile Ad Hoc Networks," *Proc. of the ACM/IEEE MOBICOM 98*, Oct. 1998.

[29] Imielinski, T., and J. C. Navas, "GPS-Based Addressing and Routing," Tech. Rep. LCSR-TR-262, CS Dept., Rutgers University, March (updated Aug.) 1996.

[30] Castaneda, R., and S. R. Das, "Query Localization Techniques for On-Demand Routing Protocols in Ad Hoc Networks," *Proc. of the 1999 ACM Mobicom Conference*, Seattle, WA, Aug. 1999.

[31] Appeleby, Steve, and S. Steward, "Mobile Software Agents for Control in Telecommunications Networks," *BT Technology Journal*, Vol. 12, No. 2, Apr. 1994, pp. 104–113.

[32] Krause, S., and T. Magedanz, "Mobile Service Agents Enabling Intelligence on Demand in Telecommunications," *Proc. of the IEEE GLOBCOM '96*, 1996.

[33] Schoonderwoerd, R., "Ant-Based Load Balancing in Telecommunications Networks," *Adaptive Behavior*, Vol. 5, No. 2, 1997, pp. 169–207.

[34] Roy Choudhury, Romit, S. Bandyopadhyay, and K. Paul, "A Distributed Mechanism for Topology Discovery in Ad Hoc Wireless Networks Using Mobile Agents," *Proc. of the First International Workshop on Ad Hoc Networks, ACM/IEEE MobiHoc 2000, in conjunction with MobiCom 2000*, Boston, MA 2000.

[35] Roy Choudhury, Romit, Krishna Paul, and S. Bandyopadhyay, "An Agent-Based Connection Management Protocol for Ad Hoc Wireless Networks," *Journal of Network and System Management*, to appear.

[36] Bambos, Nicholas, "Towards Power-Sensitive Network Architecture in Wireless Communications: Concepts, Issues and Design Aspects," *IEEE Personal Communications*, June 1998, pp. 50–59.

[37] Chang, Jae-Hwan, and L. Tassiulas, "Energy Conserving Routing in Wireless Ad Hoc Networks," in *Proc. IEEE INFOCOM 2000*, Vol. 1, 2000, pp. 22–31

[38] Singh, S., M. Woo, and C. S. Raghavendra, "Power-Aware Routing in Mobile Ad Hoc Networks," *ACM MOBICOM*, 1998, pp. 181–190

[39] Singh, S., and C. S. Raghavendra "Power Efficient MAC Protocol for Multihop Radio Networks," *Proc. of IEEE PIRMC'98 Conf.*, Vol. 1, Sept. 1998, pp. 153–157.

[40] Wu, Shih-Lin, and Yu-Chee Tseng, "Intelligent Medium Access for Mobile Ad Hoc Networks with Busy Tones and Power Control," *JSAC*, Vol. 18, No. 9, Sept. 2000.

[41] Monks, Jeffrey P., Vaduvur Bharghavan, and Wen-Mei Hwu, "A Power Controlled Multiple Access Protocol for Wireless Packet Networks," *IEEE INFOCOM 2001*, Anchorage, AK, April 2001.

[42] Paul, Krishna, and S. Bandyopadhyay, "Self-Adjusting Transmission Range Control of Mobile Hosts in Ad Hoc Wireless Networks for Stable Communication," Presented in the *Sixth International Conference on High Performance Computing HiPC99*, Calcutta, India, Dec. 1999.

[43] Ramanathan, R., and R. Rosales-Hain, "Topology Control of Multihop Wireless Networks Using Transmit Power Adjustment," *Proc. of the IEEE Conference on Computer Communications (INFOCOM)*, Tel Aviv, Israel, March 2000, pp. 404–413.

[44] Takagi, H., and L. Kleinrock, "Optimal Transmission Ranges for Randomly Distributed Packet Radio Terminals," *IEEE Trans. Commun.*, Vol. COM-32, pp. 246–257, March 1984.

[45] Cheng, Y. C., and T. G. Robertazzi, "Critical Connectivity Phenomena in Multihop Radio Network," *IEEE Trans. Commun.*, 37, 1989, pp. 770–777.

[46] Philips, T., S. Panwar and A. Tantawi, "Connectivity Properties of a Packet Radio Network Model," *IEEE Transaction on Information Theory*, Vol. 35, No. 5, Sept. 1989.

[47] Gupta, P., and P. R. Kumar, "Critical Power for Asymptotic Connectivity in Wireless Networks," *Stochastic Analysis, Control, Optimization and Applications*, W. M. McEneany, G. Yin, and Q. Zhang, (eds.), Boston, MA: Birkhauser, 1998, pp. 547–566.

[48] Kirousis, L. M., et al., "Power Consumption in Packet Radio Networks," *Theoretical Computer Science 243* (2000), pp. 289–305.

[49] Sanchez, M., P. Manzoni, and Z. Haas, "Determination of Critical Transmission Range in Ad Hoc Networks," *IEEE Workshop on Multiaccess, Mobility and Teletraffic for Wireless Communications*, Oct. 1999.

[50] Royer, Elizabeth M., and Charles E. Perkins, "Transmission Range Effects on AODV Multicast Communication," to appear in *ACM Mobile Networks and Applications special issue on Multipoint Communication in Wireless Mobile Networks*.

[51] Lee, S., "Determining Transmission Range in Ad Hoc Networks", *Technical Report*, Dept. of Computer Science, University of Maryland, College Park, MD, May 2001.

[52] Chen, S., N. Bambos, and G. Pottie, "On Power Control with Active Link Quality Protection in Wireless Communications Networks," *IEEE Proc. 28th Annual Conf. on Information Sciences and Systems*, 1994.

[53] Haas, Z. J., "The Zone Routing Protocol (ZRP) for Ad Hoc Networks," Internet Draft, Nov. 1997.

[54] Das, B., R. Sivakumar, and V. Bharghavan, "Routing in Ad Hoc Networks Using a Virtual Backbone," IEEE IC3N'97, Sept. 1997, pp. 1–20.

[55] Qayyum, Amir, Philippe Jacquet, and Paul Muhlethaler, "Optimized Link-State Routing Protocol," http://www.ietf.org/ internet-drafts/ draft-ietf-manet-olsr-01.txt, Feb. 2000. IETF Internet Draft (work in progress).

[56] Aggelou, G, and Rahim Tafazolli. "Relative Distance Micro-Discovery Ad Hoc Routing (RDMAR) Protocol," http://www.ietf.org/ internet-drafts/ draft-ietf-manet-rdmar-01.txt, Sept. 1999, IETF Internet Draft (work in progress).

[57] Obraczka, K., and G. Tsudik, "Multicast Routing Issues in Ad Hoc Networks," *Proc. of the IEEE ICUPC '98*, Oct. 1998.

[58] Deering, S.E., and D. R. Cheriton, "Multicast Routing in Datagram Internetworks and Extended LANs," *ACM Transactions on Computer Systems*, Vol. 8, No. 2, May 1990, pp. 85–110.

[59] Waitzman, D., S. Deering, and C. Partridge, "Distance Vector Multicast Routing Protocol," RFC 1075, Nov. 1988.

[60] Moy, J., "Multicast Routing Extensions for OSPF," *Communications of the ACM*, Vol. 37, No. 8, Aug. 1994, pp. 61–66, 114.

[61] Ballardie, T., P. Francis, and J. Crowcroft, "Core Based Trees (CBT)—An Architecture for Scalable InterDomain Multicast Routing," *Proc. of ACM SIGCOMM'93*, San Francisco, CA, Oct. 1993, pp. 85–95.

[62] Deering, S., et al., "The PIM Architecture for Wide-Area Multicast Routing," *IEEE/ACM Transactions on Networking*, Vol. 4, No. 2, Apr. 1996, pp. 153–162.

[63] Lee, Sung-Ju, et al., "A Performance Comparison Study of Ad Hoc Wireless Multicast Protocols," *Proc. of the IEEE INFOCOM 2000*, Tel Aviv, March 26–30, 2000.

[64]   Chiang, Ching-Chuan, Mario Gerla, and Lixia Zhang, "Forwarding Group Multicast Pro-
       tocol (FGMP) for Multihop, Mobile Wireless Networks," *ACM-Baltzer Journal of Cluster
       Computing: Special Issue on Mobile Computing*, Vol. 1, No. 2, 1998

[65]   Liu, M., et al., "AMRoute: Ad Hoc Multicast Routing Protocol," CSHCN TR 99–1,
       University of Maryland, College Park, MD.

[66]   Garcia-Luna-Aceves, J. J., and E. Lmadruga, "A Multicast Routing Protocol for Ad Hoc
       Networks," *Proc. of the IEEE INFOCOM'99*, New York, March 1999.

[67]   Wu, C. W., Y. C. Tay, and C. K. Toh, "Ad Hoc Multicast Routing Protocol Utilizing
       Increasing ID-Numbers," Internet-draft, draft-ietf-manet-amris-spec-00.txt, Nov. 1998.

[68]   Lee, Sung-Ju, W. Su, and M. Gerla, "On-Demand Multicast Routing Protocol
       (ODMRP) for Ad Hoc Networks," Internet-draft, draft-ietf-manet-odmrp-01.txt, June
       1999.

[69]   Ko, Y. B., and N. H. Vaidya, "Location-Based Multicast in Mobile Ad Hoc Networks,"
       TR98-018, Deptartment of Computer Science, Texas A&M University.

[70]   Broch , J., et al., "A Performance Comparison of Multi-Hop Wireless Ad Hoc Network
       Routing Protocols," *Proc. ACM/IEEE Mobile Comput. and Network.*, Dallas, TX, Oct.
       1998.

[71]   Das, S. R., et al., "Comparative Performance Evaluation of Routing Protocols for Mobile,
       Ad Hoc Networks," *Proc. of IEEE IC3N'98*, Lafayette, LA, Oct. 1998, pp. 153–161.

[72]   Johansson, P., et al., "Scenario-Based Performance Analysis of Routing Protocols for
       Mobile Ad Hoc Networks," *Proc. of ACM/IEEE MOBICOM'99*, Seattle, WA, Aug. 1999,
       pp. 195–206.

[73]   Lee, S. J., M. Gerla, and C. K. Toh, "A Simulation Study of Table-Driven And On-
       Demand Routing Protocols for Mobile Ad Hoc Networks," *IEEE Network*, Vol. 13,
       No. 4, July 1999, pp. 48–54.

Part III
Future Issues

8

Routing in Next-Generation Wireless Networks

8.1 UMTS All-IP Networks

At the end of 1999, work started in 3GPP toward an all-IP architecture. This evolution was driven by two objectives:

1. Independence of the transport and control layer to ease the implementation of new applications up to the mobile terminal;
2. Operation and maintenance optimization for the access network.

This evolution has an impact on different parts of the network. Basically, three main (and independent) evolutions are part of this evolution toward an all-IP architecture [1]:

- Evolution moves toward a next-generation network (NGN) type of architecture in the circuit switch domain, where the MSC function is split into a control plane part (MSC server) and a user plane part (media gateway). The introduction of packet transport (IP or ATM since the network architecture is independent of the underlying transport layer) on the network subsystem (NSS) backbone also allows moving the transcoder toward the border of the public land mobile network (PLMN). As such transcoder-free operation (TrFO) is possible, which results in much better voice quality. This feature is part of release 4.

- Addition of an IP-based multimedia subsystem (IMS) that introduces the capabilities to support IP-based multimedia services, such as voice over IP (VoIP) and multimedia over IP (MMoIP), and makes use of the packet-switched network for the transport of control and user plane data. The PS domain also deals with all the mobility (handover) aspects. This feature is part of release 5.

- Introduction of IP transport technology within the UTRAN is an alternative to the ATM-based UTRAN. This feature is also part of release 5. It is important to stress that these evolutions are independent of each other and can also be deployed in a fully independent way. This means that for each of these evolutions the operator can make an independent yes or no decision.

The introduction of the IMS is driven by the demand to offer more and enhanced services to end users. It is clear that IP plays a major role in the quick introduction of new services. Operators are faced with the challenge of finding their role in these new business opportunities. Based on their specific strengths such as expertise in communication and powerful billing systems, network operators are now moving into communication-oriented services. To do this, network infrastructures must be transformed into secure, open, and flexible platforms on which third-party developers and service providers can add generic as well as customized applications rapidly and cost effectively.

The main drivers leading this transformation of the value model of the public telecom market can be analyzed along three directions:

1. *Provision of user-centric solutions:* Market demand for services are evolving from a standardized one-size-fits-all service to a fully customized service that adapts to the user's choice and preference, as well as terminal and location.

2. *Usage of the new capabilities of networks and terminals:* Available bandwidth is increasing, which enables the deployment of media-rich services making use of all capabilities for interactivity. Control capability for network resources is enabling provision of the adapted media to the user with the expected levels of QoS and security.

3. *Evolution of marketplace and business:* With the shift in value chain, new roles are defined and new stakeholders are taking their place in the market (service providers, service retailers). In parallel, deregulation and the need for timely coping with fierce competition fuel openness toward third parties and support for open service creation and provision.

Figure 8.1 shows [1] the proposed 3GPP all-IP UMTS core network architecture [2]. New elements in this architecture are:

- *Call state control function (CSCF):* The CSCF is a SIP server that provides or controls multimedia services for packet-switched (IP) terminals, both mobile and fixed.

- *Media gateway (MG):* All calls coming from the PSTN are translated to VoIP calls for transport in the UMTS core network. This media gateway is controlled by the MGCF using the H.248 protocol [2].

- *Media gateway control function (MGCF):* The first task of the MGCF is to control the media gateways via the Media Gateway Control Protocol H.248. Also, the MGCF performs translation at the call control signaling level between ISUP signaling, used in the PSTN, and SIP signaling, used in the UMTS multimedia domain.

- *Home subscriber server (HSS):* The HSS is the extension of the HLR database with subscribers' multimedia profile data.

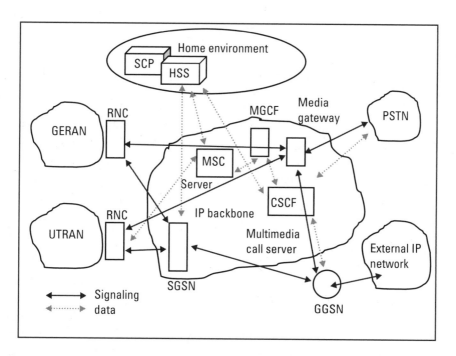

Figure 8.1 3GPP all-IP architecture (*Source:* [1]).

8.2   Routing in Distributed Wireless Sensor Networks

8.2.1   Introduction

Over the last few years, the design of micropower wireless sensor systems has gained increasing importance for a variety of applications. We have observed major advances in the design of ultra-low power sensor systems based on infra-red, vibration and acoustic sensors. They combine micropower sensor technology with a low power sensor interface, signal processing and weak-inversion RF circuits to implement entire sensor systems with the capability of forming wireless networks. Collaboration is essential between sensor nodes for the fusion of data, and the organization of the nodes can be ad hoc. While efficient protocols have been proposed to deal with ad hoc networks, large-scale wireless networks with thousands of nodes are not usually considered in that research. Moreover, energy has not been explicitly used as a metric. The availability of energy from a battery or a large power source is not always a good assumption. For example, the monitoring devices can be embedded into objects where replacing the battery is very inconvenient or impossible. The short life span of sensors and their redeployment make the topology of sensor networks quite dynamic, and these issues determine the chassis of adaptive approaches towards the design of routing protocols of such large-scale networks [3].

8.2.2   Sensor Networks

Integrated low-power sensing devices will permit remote object monitoring and tracking in many different contexts: in the field (vehicles, equipment, personnel), the office building (projectors, furniture, books, people), the hospital ward (syringes, bandages) and the factory floor (motors, small robotic devices). Networking these sensors using their wireless communication capabilities, empowering them with the ability to coordinate among themselves on a larger sensing task will revolutionize information gathering and processing in many situations. Large-scale, dynamically changing, and robust sensor colonies can be deployed in inhospitable physical environments such as remote geographic regions or toxic urban locations. They will also enable low maintenance sensing in more benign, but less accessible, environments, such as large industrial plants, and aircraft interiors.

To motivate the challenges in designing these sensor networks, consider the following scenario. Several thousand sensors are rapidly deployed (e.g., thrown from an aircraft) in remote terrain. The sensors coordinate to establish a communication network, divide the task of mapping and monitoring terrain among themselves in an energy-efficient manner, adapt their overall sensing accuracy to the remaining total resources, and reorganize upon sensor failure. When additional sensors are added or old sensors fail, the sensors reorganize themselves to take advantage of the added system resources. Several aspects of this scenario present systems design challenges different from those posed by existing computer networks. The sheer number of these devices, and their unattended deployment, will preclude reliance on broadcast communication or the configuration currently needed to deploy and operate networked devices. Devices may be battery constrained or subject to hostile environments, so individual device failure will be a common event. In addition, the configuration devices will frequently change in terms of position, reachability, power availability, and even task details. Finally, because these devices interact with the physical environment, they, and the network as a whole, will experience a significant range of task dynamics [3].

In sensor networks, most traffic is in the form of many-to-one data flows between all the networked sensors and the monitoring node. Flooding, which is normally used for disseminating information, can also be used for the network or topology discovery. The monitoring node(s) initiate flooding by passing a message to all its neighbors, each successive node forwarding it once. As flooding proceeds, nodes keep track of whom they received messages from. In case a node receives several copies of the same message from its neighbors it employs a parent selection algorithm to choose one of these as its parent. It then sends back information to its parent along the reverse direction of the original flooding message, when it needs to forward data back to a monitoring node. The network topology that is thus discovered is a spanning tree. It is of interest to know whether the discovered tree is bushy, with a few large clusters or sparse, with many smaller clusters.

In a network of thousands of nodes, it is unlikely that the designer will determine the position of each node. In an extreme case, nodes may be dropped from the air and scatter about an unknown area. To process sensor data, however, it is imperative to know where the data is coming from. Nodes could be equipped with a GPS to provide them with knowledge of their absolute position, but this is currently a costly (in volume, money, and power consumption) solution. Instead, positional information can be inferred from connection-imposed proximity constraints. In this model, only a few nodes have known positions (perhaps equipped with GPS or placed deliberately) and the remainder of the node positions are computed from knowledge about communication links [4].

8.2.3 Topology Maintenance and Sensor Deployment

Sheer numbers of inaccessible and unattended sensor nodes, which are prone to frequent failures, make topology maintenance challenging task. Hundreds to several thousands of nodes are deployed throughout the sensor field. They are deployed within tens of feet of each. The node densities may be as high as 20 nodes/m³. Deploying high number of nodes densely requires careful handling of topology maintenance. We examine issues related to topology maintenance and change in three phases [3]:

1. *Predeployment and deployment phase:* Sensor nodes can be either thrown in mass or placed one by one in the sensor field. They can be deployed by dropping from plane, delivering in an artillery shell, rocket or missile, throwing by catapult (i.e., from shipboard), placing in factory, and placing one by one either by human or robot. Although the sheer number of sensors and their unattended deployment usually preclude placing them according to a carefully engineered deployment plan, the schemes for initial deployment must reduce the installation cost, eliminate the need for any preorganization and preplanning, increase the flexibility of arrangement and promote self-organization and fault tolerance [3].

2. *Postdeployment phase:* After deployment, topology changes are due to changes in sensor nodes' position, reachability, available energy, malfunctioning and task details. Sensor nodes may be statically deployed. However, device failure is a regular or common event due to energy depletion or destruction. It is also possible to have sensor networks with highly mobile nodes. Besides, sensor nodes and the network experience varying task dynamics, and they may be targets for deliberate jamming. Therefore, sensor network topologies are prone to frequent changes after deployment [3].

3. *Redeployment of additional nodes phase:* Additional sensor nodes can be redeployed at any time to replace the malfunctioning nodes or due to changes in task dynamics. Addition of new nodes poses a need to reorganize the network. Coping with frequent topology changes in an ad hoc network that has myriad nodes and very stringent power consumption constraints requires special routing protocols [3].

8.2.4 Routing

Since any real ad hoc wireless network exhibits channel errors, delays, packet losses, and power and topology constraints, the sensor network design must consider these factors as well [5]. Typically after deployment, the first action for a sensor network is to determine its topology. This step is done because many of the traditional routing protocols require topological information for initialization. In order to conserve battery power and to reduce the probability of detection by hostile forces, it is better to use a reactive routing protocol. This is a protocol that determines a route only when it is required.

Another design choice is whether the network has a flat or hierarchical architecture. Both have advantages and disadvantages. The former is more survivable since it does not have a single point of failure; it also allows multiple routes between nodes. The latter provides simpler network management, and can help further reduce transmissions.

In a mobile ad hoc network, the changing network topology demands the network periodically reconfigure itself. Not only must the routing protocols be able to handle this situation, but also the media access mechanism. Link parameters, such as modulation type, amount of channel coding and transmitter power must adapt to the new configuration.

8.2.4.1 Energy Efficient Protocols

One of the following approaches can be used to select an energy efficient route [3, 6]:

- *Maximum power available route:* The route that has maximum total available power is preferred. The total available power is calculated by summing the available power of each node along the route. Based on this approach, a route is selected. However, this route may be a longer route; therefore, although it has higher total available power, it may not be power efficient.

- *Minimum energy (ME) route:* The route that consumes ME to transmit the data packets between the sink and the sensor node is the ME route.

- *Minimum hop (MH) route:* The route that makes the MH to reach the sink is preferred.

- *Maximum minimum available power node route:* The route along which the minimum available power is larger than the minimum available power of the other routes is preferred. This scheme precludes the risk of using up sensor nodes with low available power much earlier than the others because they are en route with nodes that have very high available power.

8.2.4.2 Data-Centric Protocols

In data-centric routing [3], the interest dissemination is performed to assign the sensing tasks to the sensor nodes. There are two approaches used for interest dissemination: sinks broadcast the interest, and sensor nodes broadcast an advertisement for the available data and wait for requests from the interested sinks. The data-centric routing requires attribute based naming, where the users are more interested in querying an attribute of the phenomenon, rather than querying an individual node. For instance, "the areas where the temperature is over 70°F" is a more common query than "the temperature read by a certain node."

Flooding is an old technique that can also be used for routing in sensor networks [3]. In flooding, each node receiving a data or management packet repeats it by broadcasting, unless maximum number of hops for the packet is reached or the destination of the packet is the node itself. Flooding is a reactive technique, and it does not require costly topology maintenance and complex route discovery algorithms. However, it has several deficiencies. In this mechanism, some nodes may receive duplicated messages and network traffic due to flooding increases exponentially with number of nodes in the network. Moreover, the flooding protocol does not take account of the available energy resources; an energy resource aware protocol must take this into account at all times.

A derivation of flooding is *gossiping* [7] in which nodes do not broadcast but send the incoming packets to a randomly selected neighbor. Once the neighbor node receives the data, it selects randomly another sensor node. Although this approach avoids the implosion problem by just having one copy of message at any node, it takes a long time to propagate the message to all sensor nodes.

Sensor Networks need to perform under varying stress levels, therefore the design of efficient protocols needs to be application specific. Detection of false alarm due to noise or even intentional jamming needs to be effectively addressed. Since power availability is one of the major constraints of such large-scale networks the design trade-offs should be efficient enough to accommodate feasible redeployment schemes due to large amounts of dead nodes. If such deployment schemes prove too costly, power harvesting/scavenging techniques should be employed. So far power scavenging has been only through solar sources. Thus monitoring environments where sensors are not exposed to a solar source can be a cumbersome process.

8.3   Pervasive Routing

Pervasive computing is a new paradigm in computing where there will be a convenient interface, through a new class of appliances, to relevant information, with the ability to either take action on it or get acted upon by it, whenever and wherever necessary. Thus, pervasive computing will help manage information quickly, efficiently, and effortlessly. Pervasive computing is about making our lives simpler. We are rapidly approaching an era where most consumer products contain an embedded computer and a tiny network interface (probably wireless). While the availability of ubiquitously "networked" goods is currently a novelty, it will soon be not only commonplace, but also all pervasive. So pervasive computing promises a computing infrastructure that seamlessly and ubiquitously aids users in accomplishing their tasks and that renders the actual computing devices and technology largely invisible. The basic idea behind pervasive computing is to deploy a wide variety of smart devices throughout our working and living spaces. These devices coordinate with each other to provide users with universal and immediate access to information and support users in completing their tasks using so-called pervasive network. Fortunately, this networking infrastructure, which is necessary to realize the vision of pervasive computing, is increasingly becoming a reality [8, 9].

Until today, for traditional networks, we have been using directed communication models: the destination of the message is specified at the time it is sent (in the case of multicast, this specification is not a single address, but a group address or a channel upon which the senders and receivers have previously agreed). However, in a system where we seriously expect millions of computers, and several orders of magnitude more active end points (or objects), it does not appear feasible that the sender of a message always specify its destination. Also, simply forwarding all traffic from a local domain onto a global pervasive network bus is infeasible because of scalability problems. Ideally, only those messages that exactly match the requirements of one or more subscribers, somewhere on pervasive network, should be sent on [10]. In effect, the backbone should subscribe to a set of messages from a local domain.

Given that this is the case, the current routing paradigms could, not by simple extension, support the pervasive computing scenario. So, what we may look for is content-based undirected routing (or application level routing). Undirected communication is where the sender of the messages does not specify their destination. This works by using a "pull" style, content-based selection of messages. Content-based routing is a fundamentally different paradigm for interaction between networked objects. By removing the necessity for producers to direct messages, we gain enormous flexibility in system architecture and scalability over traditional communication systems, allowing us to provide an interactive environment for pervasive computing. Undirected communication facilitates systems that are more easily extended, simpler to componentize, and contain a clearer mapping to real world interactions between objects [11].

Active application level routing provides a powerful but straightforward model for integrating digital appliances with the Web services. The overall idea is to help in writing the Web rather than help in reading it. The key missing Internet system element needed to aid data input is an application-level or data router. Application-level means that the unit of work is an image or a Web page, not, for example, an Internet packet. Web page, images, or other Web data are routed to transformation, aggregation, and storage applications based on a set of rules stored in the router. The applications that act on data can be running on completely different servers. In this way routing supports Internet services acting on input data. An Internet service consumes, produces, or transforms Internet data. An application-level router sends data from producer to consumer with possible transformation. Thus an application-level route is a service path for data. Such an application-level router needs to be active, which means the rule table can change based on the content received. By making the routing active (dependent upon the kind of data and the state of the network) dynamically composed services can be supported. In the simplest case this just means that we can put data in first and decide where to send it next. This kind of late-binding is common for human users of PCs, but is not the usual mode for routers in networks.

Active routing differs from Web page posting vaguely in the same way that Web page "surfing" differs from Web page reading. A Web content router selects servers for further data handling based on the content and the state of the router; the content can change the router state. These changes can be initiated by the content creator (the person with the camera) or they can result when the content creator requests services. Just as in the Web page reader case, magic can happen when the data flows to far off machines with data-transformation applications not envisioned when the system was created. Active application-level routing takes the Web-based communications and computing system and runs it "backwards." Web-based systems are data-centric. They use sophisticated client programs to pull hypertext pages to local CPUs. There many machine cycles are used to analyze the content before the next remote operation is performed. Application-level routing is also data-centric computing. But instead of pulling information to clients, it operates on data pushed back into the network, by digital appliances for example. Server and intermediating CPUs act to create useful information from raw client data. So the system will help create the Web pages of the future.

More sophisticated active routing supports service composition. The service composition idea is pretty much the same one used in operating system shells, where "pipes" are used to feed the standard output from one process into the standard input of another process. The application level router just works across machines with servers rather than across processes on a single machine and it understands a wider variety of standards. More advanced active routing includes delaying the data to combine it with data from other devices or data coming in later in time. This means the router coordinates the activity of devices. Active routing can also support trust management. The idea is to hold data until all required authorizations come through based on the available knowledge of the Internet services offered by data-services providers.

References

[1]   Vriendt, De Johan, "Mobile Network Evolution: A Revolution on the Move," *IEEE Communications Magazine*, Apr. 2002, pp. 104–111.

[2]   Bos L., and S. Leroy, "Towards an All-IP Based UMTS Architecture," *IEEE Network* Jan.–Feb. 2001, pp. 36–45.

[3]   Akyildiz, I. F., et al., "Wireless Sensor Networks: A Survey," *Computer Networks (Elsevier) Journal*, Vol. 38, No. 4, March 2002, pp. 393–422.

[4]   Doherty, L., "Algorithms for Position and Data Recovery in Wireless Sensor Networks," *M.S. thesis*, University of California at Berkeley, 2000.

[5]   Heinzelman, W. R., J. Kulik, and H. Balakrishnan, "Adaptive Protocols for Information Dissemination in Wireless Sensor Networks," *Proceedings of the ACM MobiCom'99*, Seattle, Washington, 1999, pp. 174–185.

[6]   Heinzelman, W. R., J. Kulik, and H. Balakrishnan, "Energy-Efficient Communication Protocol for Wireless Microsensor Networks," *IEEE Proc. of the Hawaii International Conf. on System Sciences*, Jan. 2000, pp. 1–10.

[7]   Hedetniemi, S., and A. Liestman, "A Survey of Gossiping and Broadcasting in Communication Networks," *Networks* Vol. 18, No. 4, 1988, pp. 319–349.

[8]   Saha, D., A. Mukherjee, and S. Bandyopadhyay, *Networking Infrastructure For Pervasive Computing: Enabling Technologies and Systems*, Boston, MA: Kluwer Academic Publishers, 2002.

[9]   Borriello G., "The Challenges to Invisible Computing," *IEEE Computer*, 2000, pp. 123–125.

[10]   Esler M., et al., "Next Century Challenges: Data-Centric Networking for Mobile Computing," *The Portolano Project at the Univ. of Washington. Proc. MobiCom 99*, Seattle, Washington, 1999.

[11]   Schilit B., N. Adams, and R. Want, "Context-Ware Commuting Applications," *Workshop on Mobile Computing Systems & Applications 1994*, Santa Cruz, CA, 1994.

9

Conclusion

The continuous progress in wireless technologies and the mobile computing paradigm provides clear indications that all-pervasive networks can be used to build cost-effective, easily deployable, high-speed digital connections needed to support the future Digital Economy. In particular, broadband wireless access technology provides a scalable and affordable local access solution that supports voice, Internet, videoconferencing, interactive gaming, video streaming as well as other high-speed data applications. A major advantage of wireless access is that it can be quickly and cost-effectively deployed. It is possible to quickly build scalable networks, also with initially limited resources. The technology is equally useful for those who already have an existing network but need to complement or expand their infrastructure. Thus, the telecommunications infrastructures can be implemented in a shorter period of time in comparison with an entire wireline solution.

The vision of future Pervasive Networking will be very simple, consisting of a core of backbone (probably wired) and a shell of access (invariably wireless). From the current pattern of network deployment, it is emerging as the most natural architecture. Historically, it is driven by the architecture of immensely successful cellular voice service. We may also call it as "wireless over wired." The last and first hop of access has to be wireless because of the requirement to support mobility. The backbone may be either wired (i.e., optical) or wireless (i.e., satellite). Since optical transmission is the preferred technology for high-speed, high bandwidth backbone and single-hop cellular is the common access technology, a popular implementation of this structure could be "cellular over optical." However, this does not imply that there are no other technologies for either backbone or access. It only indicates that these two technologies are the

191

dominant ones from both penetration rate and technological maturity. The two most distinctive characteristics that distinguish a Pervasive Network from other conventional types of networks, are its pervasiveness and its heterogeneity. It is an internetwork, comprising all existing and forthcoming networks of the world.

With this context in mind, we have focused on current trends and future technologies on wireless access with mobility support and investigated the issues in location management and routing techniques in mobile wireless networks. These networks have been broadly classified into two distinct categories: infrastructured (cellular) and infrastructureless (ad hoc) to create a ubiquitous communication as well as a computing environment for nomads which untether users from their information sources (i.e., "anytime, anywhere access to information, communication, and service"). While cellular networks usually involve a single-hop wireless link to reach a mobile terminal, ad hoc networks normally require a multihop wireless path from a source to a destination. To add mobility support in mobile wireless networks the mobility management covers generally two types of mobility: user mobility and terminal mobility.

A detailed description of the means and techniques for user location management in present cellular networks has been addressed in this book. A network must retain information about the location of endpoints in the network, in order to route traffic to the correct destinations. Location tracking (also referred to as mobility tracking or mobility management) is the set of mechanisms by which location information is updated in response to endpoint mobility. In location tracking, it is important to differentiate between the identifier of an endpoint (i.e., what the endpoint is called) and its address (i.e., where the endpoint is located). Mechanisms for location tracking provide a time varying mapping between the identifier and the address of each endpoint. Most location tracking mechanisms may be perceived as updating and querying a distributed database (the location database) of endpoint identifier-to-address mappings. In this context, location tracking has two components: (1) determining when and how a change in a location database entry should be initiated; and (2) organizing and maintaining the location database. In cellular networks, location tracking is only required when an endpoint moves from one cell to another. Location tracking typically consists of two operations: (1) updating (or registration), the process by which a mobile endpoint initiates a change in the location database according to its new location; and (2) finding (or paging), the process by which the network initiates a query for an endpoint's location (which may also result in an update to the location database). Most location tracking techniques use a combination of updating and finding in an effort to select the best trade-off between update overhead and delay incurred in finding. Specifically, updates are not usually sent every time an endpoint enters a new cell, but rather are sent according to a predefined strategy such that the finding operation can be restricted to a specific area. There is also a trade-off, analyzed formally between the update and paging costs. The location management methods are most adapted and widely used in current cellular networks (e.g., GSM, IS-54, IS-95), and are broadly classified into two groups. The first group includes all methods based on algorithms and network architecture, mainly on the processing capabilities of the system. The second group contains the methods based on learning processes, which require the collection of statistics on subscribers' mobility behavior, for instance. This type of method emphasizes the information capabilities of the network.

The cellular network architecture is single-hop networks with centralized control, while ad hoc networks are infrastructureless networks in which each node is a mobile router, equipped with a wireless transceiver. A message transfer in an ad hoc network would either take place between two nodes that are within the transmission range of each other or between nodes that are indirectly connected via multiple hops through some other intermediate nodes. In cellular wireless networks, there are a number of centralized entities to perform the function of coordination and control. In ad hoc networks, since there is no preexisting infrastructure, these centralized entities do not exist, so these networks require distributed algorithms to perform equivalent functions.

The dynamics of wireless ad hoc networks as a consequence of mobility and disconnection of mobile hosts pose a number of problems in designing proper routing schemes for effective communication between any source and destination. The proactive routing protocols that require knowledge of the topology of the entire network is not suitable in such a highly dynamic environment, since the topology update information needs to be propagated frequently throughout the network. On the other hand, a demand-based, reactive route discovery procedure generates large volume of control traffic and the actual data transmission is delayed until the route is determined. The medium access control protocols and routing protocols designed for ad hoc networks have been discussed extensively here.

One of the current research directions is based on seamless access and awareness among heterogeneous networks with mobility support. Seamless access refers to a situation where we are surrounded by a multitude of wirelessly interconnected embedded systems, mostly invisible and hidden in the background of our workplace, home, or outdoor environment. Now, more than ever, and at a faster pace than ever before, computing power can be found in smaller and more common devices. We can find inexpensive, ubiquitous networking technologies everywhere. Moore's Law continues to take its effect on computing power, networking speed, and storage media. Soon, computing power will be as free and accessible as the air we breathe. However, many devices and services speak their own languages and communicate with their own protocols. Awareness refers to the ability of the system to recognize and localize objects as well as people and their intentions in an implicit way. Pervasiveness along with heterogeneity and awareness, will give rise to a kind of internetworking, comprising all existing and forthcoming networks of the world.

List of Acronyms

ABR   Associativity-based routing

ACK   Acknowledgement

ADC   American Digital Cellular

AMPS   Advanced Mobile Phone System

AMRIS   Ad hoc multicast routing protocol utilizing increasing ID numbers

AMRoute   Ad Hoc Multicast Routing Protocol

AODV   Ad hoc on-demand distance vector routing

API   Application-programming interface

ARPANET   Advanced Research Project Agency NETwork

AS   Alternative strategy

AST   Angle-SINR Table

ATM   Asynchronous Transfer Mode

BLA   Boundary location area

BLR   Boundary location register

BS   Base station

BSP   Base station of paging

BSR   Basic spine routing

BTMA   Busy Tone Multiple Access

CA   Channel assignment

CAC   Call admission control

CAMP   Core-Assisted Mesh Protocol

CBR   Cluster-based routing

CBT   Core-based trees

CD   Code division

CDMA   Code division multiple access

CGSR   Cluster head-gateway switch routing

CIR   Carrier-to-interference ratio

CMR   Call-to-mobility ratio

CoA   Care of address

CPS   Cutoff priority scheme

CSCF   Call-state control function

CSMA   Carrier Sense Medium Access

CT2   Cordless Telephone 2

DAC   Digital-to-analog converter

DAG   Directed acyclic graph

DARPA   Defense Advanced Research Project Agency

DBTMA   Dual Busy Tone Multiple Access

DCA   Dynamic channel assignment

DCF   Distributed Coordination Function

DCS-1800   Digital Communication System at 1,800 MHz

DDB   Distributed database

DECT   Digital European Cordless Telephone

D-MAC   Directional medium access control

DSDV   Destination-sequenced distance vector routing protocol

DSR   Dynamic source routing

DVMRP   Distance vector multicast routing protocol

DVR   Distance vector routing

EIA/TIA   Electronics and Telecommunications Industry Associations

ESPAR   Electronically steerable passive array radiator

ETSI   European Telecommunication Standardization Institute

FAMA   Floor Acquisition Multiple Access

FCA   Fixed channel assignment

FD   Frequency division

FDMA   Frequency division multiple access

FGMP   Forwarding Group Multicast Protocol

FPLMTS   Future public land mobile telecommunications services

FSR   Fisheye state routing

GHA   Greedy Heuristic Algorithm

GPS   Global positioning system

GRPS   General Radio Packet Switching

GSM   Global System for Mobile Communication

GTT   Global title translation

HCA   Hybrid channel assignment

HCS   Hierarchical cellular system

HLR   Home location register

HPCS   Heterogeneous PCS

HSR   Hierarchical spine routing

HSS   Home subscriber service

IMS   IP-based multimedia subsystem

IMT-2000   International Mobile Telecommunications 2000

IR   Infrared

IS   Interim standard

ISM   Industrial, scientific, and medical

ITU   International Telecommunication Union

LA   Location area

LAR   Location-aided routing

LILT   Local information link-state topology

LINT   Local information no topology

LLC   Least cluster head change algorithm

LP-DDCA   Locally packing distributed DCA

LOS   Line of sight

LSP   Link-state packet

LSR   Link-state routing

LU   Location update

MAC   Medium access control

MACA   Multiple Access with Collision Avoidance

MACAW   Media Access Protocol for Wireless LANs

MAP   Mobile application part

MCBS   Multiple channel bandwidth system

MCDS   Minimum connected dominating set

ME   Minimum energy

MG   Media gateway

MH   Minimum hop

MGCF   Media gateway control function

MMoIP   Multimedia over IP

MOSPF   Multicast open shorted path first

MPR   Multipoint relay

MSC   Mobile switching center

MS   Mobile station

MT   Mobile terminal

MU   Mobile user

NGN   Next-generation network

NLST   Neighborhood-Link-State Table

ODMRP   On-demand multicast routing protocol

OLSR   Optimized link-state routing

OSPF   Open Shortest Path First

PA   Paging area

PACS   Personal access communication systems

PCMA   Power-Controlled MAC Protocol

PCSN   Personal communication services network

PDC   Personal Digital Cellular

PDE   Partial differential equation

PDF   Probability distribution function

PEDCP   Path evaluation and data communication protocol

PFP   Path finding protocol

PHS   Personal handy phone system

PHY   Physical

PIM   Protocol independent multicast

PLMN   Public land mobile network

PMAC   Power-sensitive MAC protocol

PR   Paging request

PSIP   Parallel-to-sequential intelligent paging

PSTN   Public switched telephone network

QoS   Quality of service

RD   Relative distance

RDM   Relative distance microdiscovery

RDMAR   Relative distance microdiscovery ad hoc routing

RF   Radio frequency

RRA   Resource allocation algorithm

RRM   Radio resource management

RTS/CTS   Request-to-send/clear-to-send

RUP   Reuse partitioning

SDR   Software-defined radio

SA   Service area

SCP   Signaling control point

SDC   Start-of-data communication

SINR   Signal-to-interference-and-noise ratio

SIP   Sequential intelligent polling

SIR   Signal-to-interference ratio

SNR   Signal-to-noise ratio

SP   Signaling point

SRMA   Split-Channel Reservation Multiple Access

SSAR   Signal stability-based adaptive routing

STP   Signaling transfer point

TD   Time division

TDMA Time division multiple access

TLDN Temporary local directory number

TORA Temporally-ordered routing algorithm

T-R Transmitter-receiver

TRCP Transmission range control protocol

TrFO Transcoder-free operation

TTM Time to migrate

UMTS Universal mobile telecommunication systems

VHE Virtual home environment

VLR Visitor location register

VoIP Voice over IP

WCDMA Wideband code division multiple access

WLAN Wireless local-area network

WPAN Wireless personal area network

WPBX Wireless private branch exchange

WRC World Radiocommunication Conference

WRP Wireless routing protocol

ZRP Zone routing protocol

About the Authors

Amitava Mukherjee has been a principal consultant of IBM Business Consulting Services, part of IBM Global Services (formerly PwC Consulting) in Calcutta, India, since 1995. He received his Ph.D. in computer science from Jadavpur University, Calcutta, India. He was with the Department of Electronics and Telecommunication Engineering at Jadavpur University, Calcutta, India, from 1983 to 1995. His research interests are in the area of mobile computing and communication, pervasive computing and M-commerce, optical networks, combinatorial optimization, and distributed systems. His interests also include the mathematical modeling and its applications in the fields of societal engineering and international relations. He is the author of over 65 technical papers, one monograph, and four books. He is a member of IEEE and the IEEE Communication Society.

Somprakash Bandyopadhyay is an associate professor with the MIS and Computer Science Group of the Indian Institute of Management, Calcutta. Prior to that, he was in charge of the Knowledge Management and Learning & Professional Development Group at PricewaterhouseCoopers Limited, a global management consultancy firm in Calcutta for 6 years. Besides this industrial experience, Dr. Bandyopadhyay has more than 15 years of experience in teaching, research, and software development at the Indian Institute of Technology, Kharagpur; the Indian Institute of Technology, Bombay; the Indian Institute of Management, Calcutta, Tata Institute of Fundamental Research, Bombay; and Jadavpur University, Calcutta. He has a Ph.D. in computer science from Jadavpur University (1985) and a B.Tech. in electronics and electrical communication engineering from the Indian Institute of Technology, Kharagpur

(1979). He was a fellow of the Japan Trust International Foundation and worked at the Advanced Telecommunication Research Institute in Kyoto, Japan, for 8 months in 2001, in the area of ad hoc wireless networks. He was also a fellow of the Alexander von Humboldt Foundation, Germany, and was involved in postdoctoral research at the German Research Center for Artificial Intelligence, Saarbrucken, for a year in 1989–1990.

Debashis Saha is currently an associate professor with the MIS and Computer Science Group of the Indian Institute of Management, Calcutta, India. He received his B.E. in electronics and telecommunication engineering. from Jadavpur University, Calcutta, India, in 1986, and his M.Tech. and Ph.D., both in electronics and electrical communication engineering, from the Indian Institute of Technology (IIT) at Kharagpur, India, in 1988 and 1996, respectively. He was a senior research scholar at IIT, Kharagpur, between 1988 and 1990, while conducting research on protocol engineering. He was with Jadavpur University as a faculty member in the Computer Science & Engineering. Department from 1990 to 2001. His research areas are network protocols, WDM optical networks, wireless and mobile communication and networking, and pervasive computing. He has published more than 100 papers in various conferences and journals and delivered several invited talks and tutorials in networking conferences and workshops. He is currently the principal investigator of two major funded projects on WDM optical networking research initiatives. He has coauthored four books and a monograph. Dr. Saha is a life-member of the Computer Society of India (CSI), a member of IFIP WG 6.8, a Senior Member of IEEE, and a member of the IEEE Computer Society and IEEE Communication Society. He is a recipient of the prestigious Career Award for Young Teachers (1997) from the All India Council for Technical Education (AICTE), Government of India, and is a SERC visiting fellow (1999) and a BOYSCAST fellow (2000) of the Department of Science and Technology (DST), Government of India.

Index

Recent Titles in the Artech House Mobile Communications Series

John Walker, Series Editor

For further information on these and other Artech House titles, including previously considered out-of-print books now available through our In-Print-Forever® (IPF®) program, contact:

Artech House
685 Canton Street
Norwood, MA 02062
Phone: 781-769-9750
Fax: 781-769-6334
e-mail: artech@artechhouse.com

Artech House
46 Gillingham Street
London SW1V 1AH UK
Phone: +44 (0)20 7596-8750
Fax: +44 (0)20 7630-0166
e-mail: artech-uk@artechhouse.com

Find us on the World Wide Web at:
www.artechhouse.com

The Artech House Universal Personal Communications Series

Ramjee Prasad, Series Editor

CDMA for Wireless Personal Communications, Ramjee Prasad

IP/ATM Mobile Satellite Networks, John Farserotu and Ramjee Prasad

OFDM for Wireless Multimedia Communications, Richard van Nee and Ramjee Prasad

Radio over Fiber Technologies for Mobile Communications Networks, Hamed Al-Raweshidy and Shozo Komaki, editors

Simulation and Software Radio for Mobile Communications, Hiroshi Harada and Ramjee Prasad

TDD-CDMA for Wireless Communications, Riaz Esmailzadeh and Masao Nakagawa

Technology Trends in Wireless Communications, Ramjee Prasad and Marina Ruggieri

Third Generation Mobile Communication Systems, Ramjee Prasad, Werner Mohr, and Walter Konhäuser, editors

Towards a Global 3G System: Advanced Mobile Communications in Europe, Volume 1, Ramjee Prasad, editor

Towards a Global 3G System: Advanced Mobile Communications in Europe, Volume 2, Ramjee Prasad, editor

Universal Wireless Personal Communications, Ramjee Prasad

WCDMA: Towards IP Mobility and Mobile Internet, Tero Ojanperä and Ramjee Prasad, editors

Wideband CDMA for Third Generation Mobile Communications, Tero Ojanperä and Ramjee Prasad, editors

Wireless IP and Building the Mobile Internet, Sudhir Dixit and Ramjee Prasad, editors

WLAN Systems and Wireless IP for Next Generation Communications, Neeli Prasad and Anand Prasad, editors

WLANs and WPANs towards 4G Wireless, Ramjee Prasad and Luis Muñoz